Power Electronics, Radio Frequency and Microwave Engineering

Edited by Raúl Gregor, Kim Ho Yeap and Augustine O. Nwajana

Published in London, United Kingdom

Power Electronics, Radio Frequency and Microwave Engineering
http://dx.doi.org/10.5772/intechopen.102312
Edited by Raúl Gregor, Kim Ho Yeap and Augustine O. Nwajana

Contributors
Kim Ho Yeap, Victor Oluwaferanmi Adewuyi, Salahaldein A. Rmila, Moufu Kong, Nizar Sghaier

First published in London, United Kingdom, 2023 by IntechOpen
IntechOpen is the global imprint of INTECHOPEN LIMITED, registered in England and Wales, registration number: 11086078, 5 Princes Gate Court, London, SW7 2QJ, United Kingdom

British Library Cataloguing-in-Publication Data
A catalogue record for this book is available from the British Library

Additional hard and PDF copies can be obtained from orders@intechopen.com

Power Electronics, Radio Frequency and Microwave Engineering
Edited by Raúl Gregor, Kim Ho Yeap and Augustine O. Nwajana
p. cm.
Print ISBN 978-1-80356-911-6
Online ISBN 978-1-80356-912-3
eBook (PDF) ISBN 978-1-80356-913-0

For EU product safety concerns:
IN TECH d.o.o., Prolaz Marije Krucifikse Kozulić 3, 51000 Rijeka, Croatia,
info@intechopen.com or visit our website at intechopen.com.

Meet the editors

Raúl Gregor received a degree in electronic engineering from the Catholic University of Asunción, Paraguay, in 2005. He received his MSc and Ph.D. in Electronics, Signal Processing, and Communications from the Higher Technical School of Engineering (ETSI), University of Seville, Spain, in 2008 and 2010, respectively. Since March 2010, Dr. Gregor has been the head of the Laboratory of Power and Control Systems (LSPyC), Engineering Faculty, National University of Asunción (FIUNA), Paraguay. He is currently the head of the Department of Electronic and Mechatronics Engineering at FIUNA. Dr. Gregor has authored or coauthored about 150 technical papers in the field of power electronics and control systems. He obtained the Best Paper Award from *IEEE Transactions on Industrial Electronics* in 2010 and the Best Paper Award from *IET Electric Power Applications* in 2012. His research interests include multiphase drives, advanced control of power converter topologies, power quality, renewable energies, modeling, simulation, optimization and control of power systems, smart metering and smart grids, and predictive control. Since 2017, he has led the technology-based company PQEnerSol SRL developing cutting-edge technology in the energy sector.

Kim Ho Yeap is an associate professor at Universiti Tunku Abdul Rahman, Malaysia. He is a senior member of the Institute of Electrical and Electronics Engineers (IEEE), a professional engineer registered with the Board of Engineers, Malaysia, a chartered engineer registered with the UK Engineering Council, and an Association of Southeast Asian Nations (ASEAN) chartered professional engineer. He received his BEng (Hons) in Electrical and Electronics Engineering from Universiti Teknologi Petronas in 2004, MSc in Microelectronics from Universiti Kebangsaan Malaysia in 2005, and Ph.D. from Universiti Tunku Abdul Rahman in 2011. In 2008 and 2015, respectively, he underwent research attachment at the University of Oxford, UK, and Nippon Institute of Technology, Japan. He has published more than 150 research articles, including patents, refereed journal papers, conference proceedings, books, and book chapters.

Augustine O. Nwajana holds a Ph.D. in Electrical and Electronic Engineering from the University of East London, UK, and a research CPD certificate in Practical Antenna Design from the University of Oxford, UK. He was with Siemens AG as a telecoms engineer from 2005 to 2009, where his experience spanned many countries, including the United States, United Kingdom, United Arab Emirates, Germany, South Africa, Ghana, and Nigeria. He was also with Coventry University, UK (as an associate lecturer) and with the University of East London (as a part-time lecturer). Dr. Nwajana is currently a senior lecturer at the University of Greenwich, UK, where his current research interests include radio frequency (RF) energy harvesting/scavenging, battery-less RF/microwave systems, passive RF/microwave devices, and more. He has authored/co-authored five books, several journal articles, book chapters, and conference papers. He is currently an editor for IGI Global USA and Electronics MDPI Switzerland. He has served as a keynote/Invited speaker and TPC member for several conferences. He is a senior member of the Institute of Electrical and Electronics Engineers (IEEE) and a fellow of the UK Higher Education Academy. He was a recipient of the Federal Government of Nigeria Scholarship Award.

Contents

Preface

In today's fast-evolving technological landscape, the realm of power electronics, radio frequency (RF), and microwave engineering stands as a cornerstone of modern innovation. The remarkable expansion of power converter technology has ushered in a new era, where its influence permeates through an extensive spectrum of low-, medium-, and high-power applications. The intrinsic ability to efficiently manage electrical energy has catalyzed this surge, revolutionizing industries and redefining possibilities.

Central to this transformation is the high penetration of cutting-edge microprocessors with the capacity to implement high-performance nonlinear digital controllers. These dynamic controllers, coupled with recent strides in the development of high-frequency switching power electronic devices, have ushered in a paradigm shift. The reduction in on-state loss and consequential switching loss of power semiconductors has significantly bolstered the efficiency of modern power converters. Consequently, these advancements have not only shrunk the size of power converters but also mitigated heat generation, making these devices more efficient in the energy conversion process.

Amidst this scenery, a multitude of novel power converter topologies has surfaced in recent literature, each tailored to emerging applications. These innovative designs encompass diverse criteria and intricacies, often intricately intertwined with the digital control systems they employ. This book provides readers with a comprehensive overview of the prevailing state of the art, covering topics ranging from design principles to the realm of emerging applications, and delving deep into the nuances of control systems. Our intention is to provide a coherent narrative that encapsulates recent breakthroughs in topics such as power electronics, RF, and microwaves while concurrently fostering an environment of exploration and innovation on these topics.

We extend our gratitude to the contributors who have shared their knowledge, and we are confident that this volume will serve as a useful resource for researchers, practitioners, and enthusiasts alike. It is our sincere hope that *Power Electronics, Radio Frequency and Microwave Engineering* sparks new ideas, fosters collaboration, and continues to propel the evolution of this captivating field.

Sincerely,

Raúl Gregor
Faculty of Engineering,
Department of Electronic and Mechatronic Engineering,
National University of Asunción,
San Lorenzo, Paraguay

Kim Ho Yeap
Associate Professor,
Universiti Tunku Abdul Rahman,
Petaling Jaya, Malasyia

Augustine O. Nwajana
Senior Lecturer,
University of Greenwich,
London, United Kingdom

Section 1

Power Electronics

Chapter 1

Introductory Chapter: Wireless Power Transmission – An Overview

Kim Ho Yeap

1. Introduction

Wireless power transmission (WPT) refers to the process of transferring and harvesting electrical energy without the use of wires and cables. Although this technology has only started to gain traction in about a decade or two ago, the idea of realizing WPT is not new and can be dated back to the turn of the twentieth century, about 150 years ago. Inspired by the experimental observations reported independently by André-Marie Ampère, Michael Faraday, and Carl Friedrich Gauss, James Clerk Maxwell established the unified theory of electricity and magnetism and published his formulations in his textbook "A Treatise on Electricity and Magnetism" in 1873 [1]. These four sets of notable equations outline the fundamental principles of electromagnetism and have contributed significantly toward the development of technology based on electromagnetic fields and waves. In the late 1800, Heinrich Hertz conducted a series of experiments, which successfully verified the presence of electromagnetic waves theorized by Maxwell. In one of his experiments, Hertz showed the propagation of waves in air. In 1886 to 1889, Hertz proved that the electric and magnetic field vectors are perpendicular to each other and to the propagating direction [1]. The discoveries of these scientists have laid the cornerstone that sparked the exploration of wireless technology. Nikola Tesla was the first scientist to demonstrate experimentally the feasibility of transferring power wirelessly *via* inductive and capacitive couplings. In 1881, Tesla invented the Tesla coil, *viz.* a resonant transformer with the coils separated by air-gap, which could be used to generate low-current, high-voltage, and high-frequency electricity. By connecting the Tesla coil to a coupled oscillatory circuit with a spark gap driven by a low-frequency alternating-current (AC) or direct current (DC) source, he showed that light could be generated by the strong electric field when it was passing through a rarefied tube [2]. Tesla's continuous relentless study on this matter gave him the epiphany that the technology could one day be used to drive motors and switch lamps at considerable distance from the source. Following this discovery, Tesla gave a series of lectures and demonstrations on his wireless technology in 1892 and 1893 and among the gist of his lectures was his speculation of using wireless technology to transmit information and power. Wireless telecommunication and wireless power transmission are ubiquitous to mankind today; however, these concepts were flabbergasting when it was first suggested by Tesla more than a century ago. Tesla lit light bulbs across the stage when demonstrating before the American Institute of Electrical Engineers and at the 1893 Columbian Exposition in Chicago [3]. In 1899, he used a 108 V high-frequency power to wirelessly turn on Geissler tubes [4, 5]. He further showed in his Colorado Springs laboratory in 1899 to 1900 that his experimental setup was able to lit three incandescent lamps at a distance of about 30.5 m [3]. In March 1901, Tesla obtained $150,000 from J. P. Morgan to build the 57-m tall Wardenclyffe tower in Shoreham, New York. The Wardenclyffe tower was intended to be used as a power plant to transfer electrical energy through the ionosphere, over large distances. It was indeed unfortunate enough, though, the project ended in vain in 1905. Tesla had no choice but to abandon it when he failed to secure investments to support the Wardenclyffe tower project. Due to his bold idea on wireless power transmission at his time and his discovery on wireless communication, nevertheless, Nikola Tesla was regarded as the "Father of Wireless Technology" today [6].

The daily lives of mankind today are interwoven seamlessly with electrical and electronic (E&E) devices. Wherever we are and whatever we are doing, there are certainly E&E devices that we rely on. This is to say that E&E products have become an indispensable commodity to mankind. The variability of these E&E products is wide and may range from telecommunication gadgets such as cell phones, to living accessories such as toothbrushes and razors, to transports such as electric cars and scooters. It is to be noted that E&E devices could only operate with the supply of electrical power. One way to empower these devices is to tether them to the power grid. Although it is immaculately fine to do so, fettering an E&E device with wires actually restricts its portability and mobility, rendering it cumbersome to be used and limiting the distance of its usage. Take for instance, a vacuum cleaner or a phone that can only operate when it is connected to the source using wires. The functional distance of the device is significantly bounded by the effective length of the wire. The inconvenience caused by wires can be overcome if the devices are modified to operate wirelessly, using batteries. The advent of the wireless technology provides the flexibility for users to use the E&E devices remotely, without the worries of the limitation imposed by distance. Even so, however, a wireless device such as a cellular phone or an unmanned aerial vehicle (UAV) is, strictly speaking, not entirely "wireless" after all. Despite the prevailing advancement of wireless technology, there still comes a time where a wireless device has to be connected by wires to the power grid for battery recharging. An E&E device could only become completely portable and mobile when it is disentangled from this last cut of wires. To do so, the batteries attached to the device have to be facilitated with the function of being wirelessly charged. The hope to realize a fully wireless E&E device has prompted the scientific community to devote vigorous researches related to wireless power transmission [7]. This chapter presents a brief overview of different wireless power transmission (WPT) methods. Like a wireless telecommunication system, the energy radiated from a WPT system attenuates along with distance. To improve the effective distance of power transmission, parasitic wires can be integrated into the system. An elaboration of this method is described at the later part of the chapter.

2. Wireless power transmission mechanisms

Based on the separation distance d between the transmitting and receiving antennas, the mechanisms are employed to transfer power wirelessly bifurcate into those used in the (i) near- and (ii) far-field regions. As can be seen in **Figure 1**, when both the antennas are placed in close proximity, within the distance of a wavelength (λ), the WPT system is said to be operating in the near-field region. In this region, both reactive and radiative fields exist. Fields close to the transmitting antenna (i.e., $d \leq \frac{\lambda}{2\pi} \approx 0.16\lambda$) are literally reactive and the phases between the

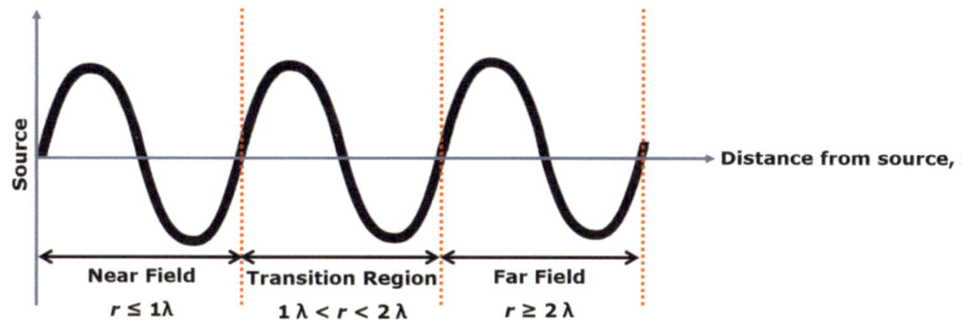

Figure 1.
Field regions for antenna size not more than 0.5λ.

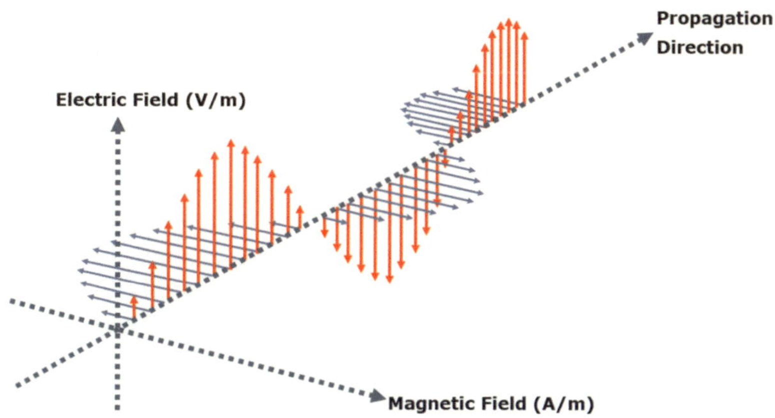

Figure 2.
Transverse electromagnetic wave.

electric and magnetic fields differ by 90°. In the reactive near field, the energy exchanges periodically between the reactive loads and the source, resulting in zero average power. At $\frac{\lambda}{2\pi} \le d \le \lambda$, and the fields start to radiate; but the intensity of the fields decays at a fast rate, which is inversely proportional to d^3 and, likewise, the power density is inversely proportional to d^6. When the distance between both antennas is at least two wavelengths (2λ) apart, the system is said to operate in the far field. Owing to propagation delay, the electric and magnetic fields are orthogonal, but in phase with each other in this region, both fields polarize perpendicularly to the direction of propagation. This type of wave is known as the transverse electromagnetic or TEM wave and is graphically depicted in **Figure 2** [8]. At the far-field region, the fields decay as $\frac{1}{d}$, which is relatively slower than the case of the radiative near field, while the power density decays as $\frac{1}{d^2}$. As can be seen in **Figure 1**, the transition zone lies between the near and far fields. Both near and far fields fade into each other here. Hence, the transition zone exhibits the characteristics of both fields in this region.

2.1 The near-field mechanism

The near-field mechanism transmits power wirelessly from the source to the load based on the inductive or capacitive coupling. In inductive or magnetic

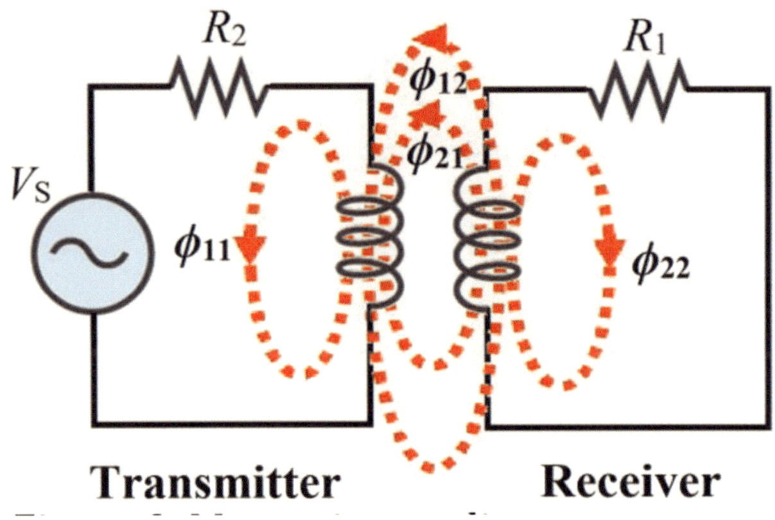

Figure 3.
Magnetic coupling.

coupling, power is generated at the load *via* the variation of magnetic flux. The operational concept of this type of energy transfer is rather similar to how a transformer works. **Figure 3** depicts the circuit schematic used for inductive coupling. As can be seen in figure, the AC source at the transmitter creates magnetic flux at the coils. The total magnetic flux can be classified into two parts—the individual flux induced in the transmitter and receiver coils and the flux induced by the transmitter coil that interlinks to the receiver coil, and *vice versa*. According to Faraday's law, when the magnetic flux density **B** varies with time t, voltage V is induced, that is, [8]

$$V = -\int_S \frac{\partial \mathbf{B}}{\partial t} \cdot dS \tag{1}$$

where S is an arbitrary surface in space. Hence, by virtue of this law, power is delivered from the transmitter source to the receiver load, without the physical need of connecting the transmitter and receiver circuits together [6]. When power is transferred from the primary coil to the secondary, a fraction of it tends to be lost in air. To minimize the loss, it is advisable to place the circuits at the same plane and at close proximity. The coupling between the coils can also be increased using wide and flat coils. Like the case of a transformer, the intensity of the magnetic field and current is directly proportional to the number of turns of the coil n. Hence, by increasing n, the efficiency of inductive coupling can be improved accordingly. Some of the available WPT systems that work based on the inductive coupling concept are rechargeable toothbrushes and razors.

Since the fields in this region are non-radiative in nature, the energy decays rapidly. The effective distance of the power transferred using the inductive coupling method is not more than the size of the antenna. To increase the effective distance, the transmitter and receiver circuits can be modified into resonant circuits. This type of WPT method is known as resonant inductive coupling (RIC). To enable power transmission, the primary and secondary coils have to resonate at the same frequency. The resonance frequency f_r can be determined using Eq. (2) below

$$f_r = \frac{1}{2\pi\sqrt{LC}} \tag{2}$$

where L is the inductance of the coil and C the capacitance in the circuits. By employing RIC, the effective distance can be extended by 10 times of that obtained by the conventional inductive coupling method. Charging pads for cell phones and other handheld devices are examples of WPT systems, which operate using RIC. Some implantable medical devices (IMDs), such as the implantable glucose sensors used mostly by diabetes patients and pacemakers for patients with irregular heart palpitations, are remotely charged using the RIC method as well.

The capacitive or electric coupling method makes use of the variation electric flux to generate power at the load. **Figure 4** illustrates the different types of circuit configurations used for capacitive coupling [9, 10]. Upon inspection on the figure, it can be observed that pairs of electrodes are commonly found in all configurations. The electrodes are usually made of metallic plates. With air sandwiched in between the plates, capacitors are formed. According to Maxwell, the time rate of change of electric flux density **D** leads to a displacement current I_D [11], that is,

$$I_D = \frac{\partial \mathbf{D}}{\partial t} \tag{3}$$

Hence, when the electric flux imposed on the coupled plate varies with time (due its connection to the oscillating voltage source), a displacement current is generated to deliver power from the transmitter to the receiver [12]. Since the plates act as capacitors, the efficiency of power transfer can be enhanced by increasing the area of the plates and reducing the intervening air gap. Some charging pads and IMDs also operate based on capacitive coupling. Like the case of its inductive counterpart, the effective distance of capacitive coupling is rather limited. Hence, resonant circuits can also be introduced into it to ameliorate its performance. **Figure 5** depicts the schematic of a resonant capacitive coupling circuit [13]. The transmitter circuit in the figure is supplied with a V_{in} source with an internal resistance of R_{in} embedded in it and the energy generated is to be transferred to the R_{load} resistor at the receiver circuit. The transmitter and the receiver circuits are capacitively coupled through capacitances C_3 and C_4. Capacitor C_1 and inductor L_1 are incorporated into the transmitter circuit to generate the resonance effect. Likewise, C_2 and L_2 serves the same purpose as C_1 and L_1 at the receiver circuit. The resistors R_1 and R_2 account for the losses experienced, respectively, by L_1 and L_2.

2.2 The far-field mechanism

To realize wireless power transmission in the far-field region, the energy carried by the electromagnetic waves has to be radiated from the transmitting antenna, propagates in air for a distance, and is finally coupled to the receiving antenna.

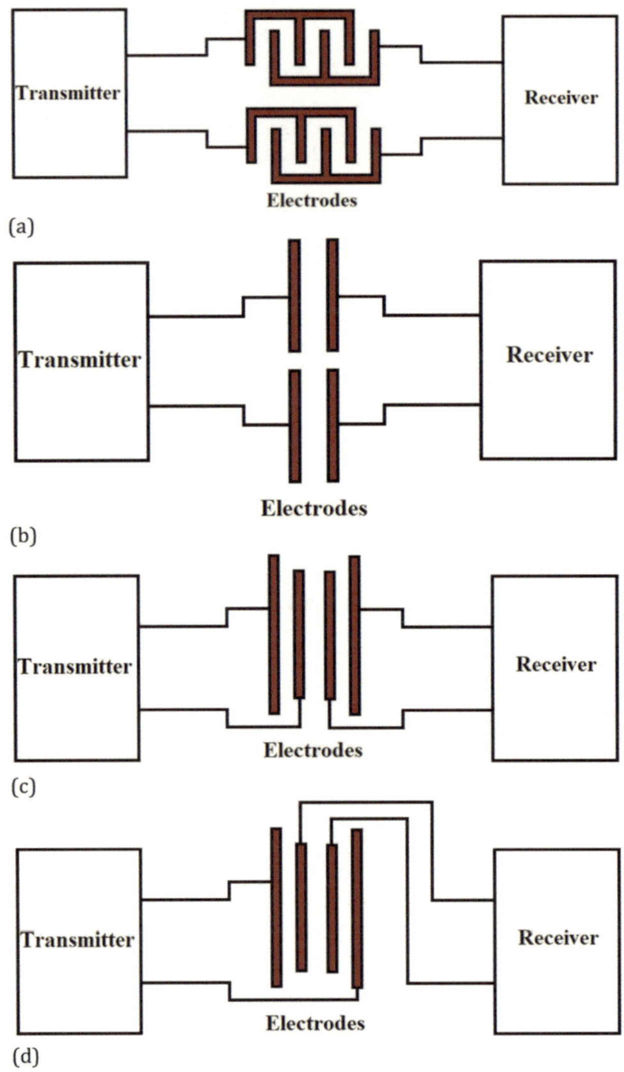

Figure 4.
Different types of capacitive coupling structures, that is, the (a) stack array structure, (b) the conventional two-plate structure, and the modified two-plate structures proposed by (c) Zhang et al. *[9] and, (d) Han* et al. *[10].*

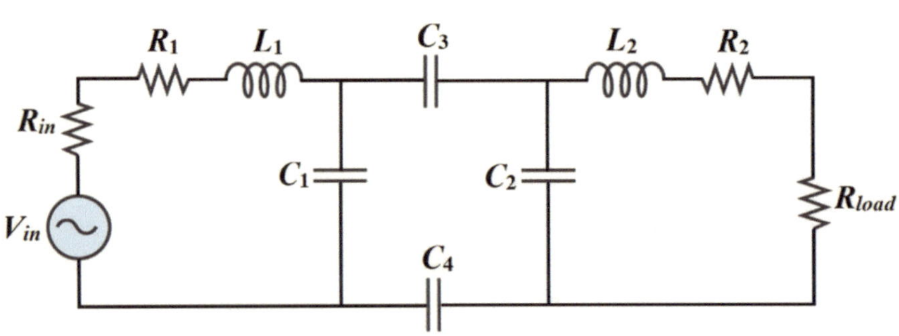

Figure 5.
Circuit model of a resonant capacitive coupling circuit.

For an electromagnetic wave to propagate, it has to satisfy Helmholtz's equation. Since air is lossy and homogeneous in nature and assuming that it is also source free, Helmholtz's equations for the electric **E** and magnetic **H** fields can be written as

$$\nabla^2 \mathbf{E} + k^2 \mathbf{E} = 0 \tag{4}$$

$$\nabla^2 \mathbf{H} + k^2 \mathbf{H} = 0 \tag{5}$$

where $k = \beta - j\alpha$ is the wavenumber in air and β and α are, respectively, the phase and attenuation constants, which, for a TEM wave, can be expressed as [14],

$$\alpha = \omega \sqrt{\frac{\mu\varepsilon}{2}\left[\sqrt{1 + \left(\frac{\sigma}{\omega\varepsilon}\right)^2} - 1\right]} \tag{6}$$

$$\beta = \omega \sqrt{\frac{\mu\varepsilon}{2}\left[\sqrt{1 + \left(\frac{\sigma}{\omega\varepsilon}\right)^2} + 1\right]} \tag{7}$$

where ω is the angular frequency, and σ, ε and μ are, respectively, the electrical conductivity, electrical permittivity, and magnetic permeability of air. It is worthwhile noting that air is usually assumed lossless in most textbooks because its conductivity is very small. In reality, however, the conductivity of air ranges from 3 to 8 fS/m, depending on the humidity of air [15]. The solutions of (4) and (5) represent waves propagating with a velocity of $\frac{1}{\sqrt{\mu\varepsilon}} \approx 3 \times 10^8 \mathrm{m/s}$.

The efficiency of wireless power transmission is essentially determined by the designs of the transmitting and receiving antennas. To ensure that the signal is effectively delivered by the transmitting antenna, the following performance indicators are used for design assessments [14]:

i. Radiation pattern

   The radiation or antenna pattern is a graphical representation of the field intensity or power density in terms of spatial distribution, when the field radiated from the antenna.

ii. Radiation intensity

   The radiation intensity refers to the power radiated by the antenna at a unit solid angle.

iii. Directive gain

   The directive gain shows the concentration of the radiated power at a specific direction. It compares the radiation intensity in that direction with the average radiation intensity.

Since waves tend to diffract as they propagate, spill over loss is inevitable. To minimize such loss, high-directivity antennas capable of exhibiting radiation patterns in the shape of a pencil beam are preferred.

When collecting the power from the radiation, the effective area of the receiving antenna is to be optimized. The effective area dictates the ability of the receiving antenna in coupling energy scattered to it. The effective area A_e is given as

$$A_e = \frac{G_d \lambda^2}{4\pi} \quad (8)$$

where G_d denotes the directive gain. In 1946, the Danish-American engineer, Harald Trap Friis modified the effective area equation in (8) so as to relate the power received by the receiving antenna P_r from the transmitting antenna P_t at far field. The equation, which is more commonly called the Friis transmission formula these days, is mathematically described in (9) below,

$$P_r = G_{dr} G_{dt} \left(\frac{3 \times 10^8}{4\pi d f} \right)^2 P_t \quad (9)$$

where f is the operating frequency and G_{dr} and G_{dt} are, respectively, the directive gains of the receiving and transmitting antennas. Friis formula comes in handy when designing the receiving antenna since the relationship between both the transmitting and receiving antennas can be easily found, without the need to figuring out the effective area. Upon inspection of (9), it can also be observed that the received power P_r drops with the square of the signal frequency f. This is to say that, although signal with higher frequency has the advantages of carrying higher energy and that the size of the system that supports it can be greatly miniaturized, it suffers the drawback of high path loss. The loss is particularly conspicuous when the distance of energy transfer is far. Hence, for a system to transfer energy wirelessly to a reasonably long distance, Rosa et al. [16] recommended f = 900 MHz to be the best trade-off.

Since both the WPT and telecommunication systems radiate and collect electromagnetic signal, their theory of operations is rather similar. **Figure 6** depicts the block diagram of a general WPT system. As can be seen from the transmitter module, the AC source is first rectified to produce DC signal. The signal is then modulated to radio frequencies (RF) and emitted to air. The RF wave is harvested by the rectenna at the

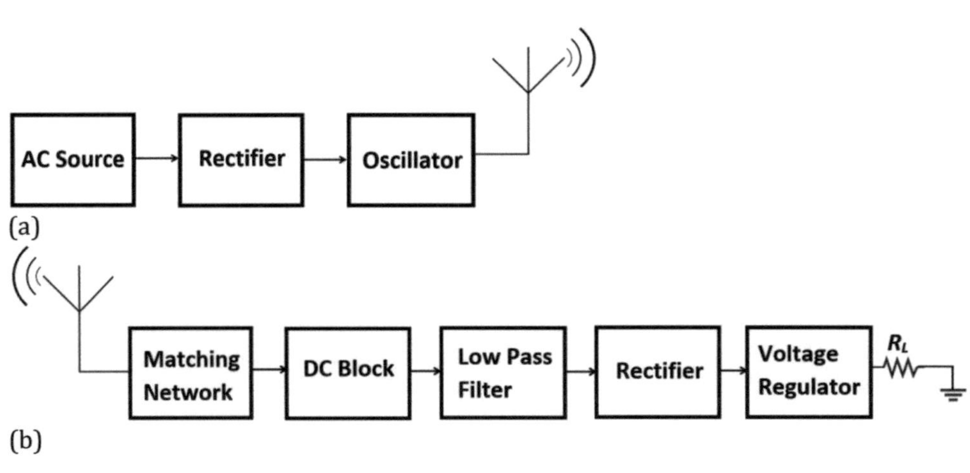

Figure 6.
Block diagrams for the (a) transmitter and (b) receiver circuits.

receiver. The rectenna constitutes the receiving antenna and a rectifier circuit, which is used to convert the RF signal into DC power. As can be seen in **Figure 6**, the output of the rectenna is connected to the load resistor. Upon coupling the incident RF signal to the receiving antenna, the energy is fed to a matching network such as an open-ended single stub tuner [17]. The purpose of the matching network is to provide impedance matching, so that wave reflections can be minimized [18]. A DC block is connected between the matching network and the rectifier circuit to filter any DC signal from passing to the rectifier circuit. The rectifier circuit constitutes nonlinear devices such as diodes. Various spectral components are produced at the output of the nonlinear devices, including the DC component, the AC fundamental frequency and its harmonics, and the intermodulation mixing products [19]. The low-pass filter connected in cascade with the rectifier circuit is used to filter the unwanted components, so that only the DC signal remains, and to recover the efficiency degradation caused by the junction capacitance of the diodes [17]. The last block before the DC signal reaches the load resistor R_L is the regulator, which can be easily constructed using a parallel-connected capacitor and Zener diode. The regulator suppresses the ripples so that a smoothened DC signal is received at the output.

3. Optimization using parasitic wires

As mentioned in the preceding section, the power density of the waves attenuates rapidly along with the propagation distance. This phenomenon is corroborated by the experimental findings by Kurs et al. in [20]. When sending a 9.9 MHz signal to remotely light up a light bulb, the researchers found that the efficiency of the RIC system dropped more than 50% at a distance above 2 m, which corresponds to about 0.066λ. The result will, of course, be worse for a nonresonant WPT system.

One way to improve the effective distance of wireless power transmission is to introduce parasitic wires into the system [7, 21, 22]. As shown in **Figure 7**, the parasitic wire, which could be in the form of circular or rectangular geometry, is inserted in between and along the same axis as the transmitting and receiving

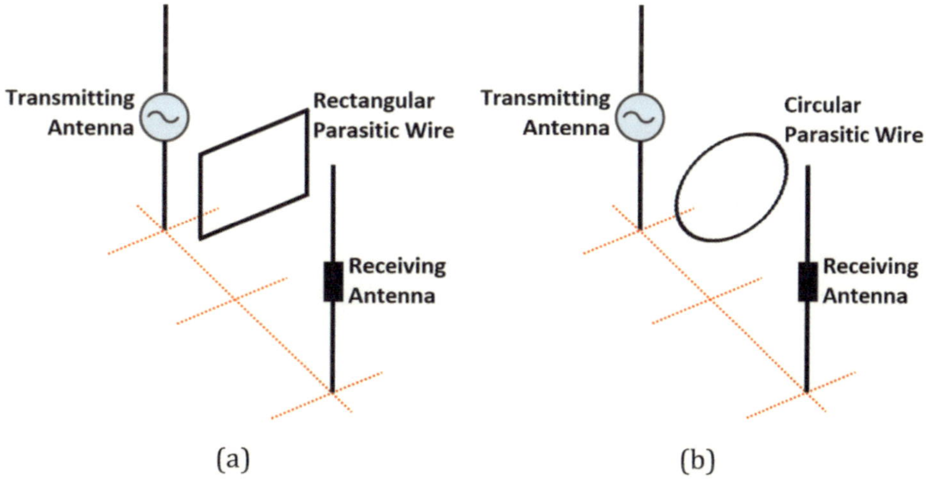

Figure 7.
A wireless power transmission system integrated with a (a) rectangular and a (b) circular parasitic wire.

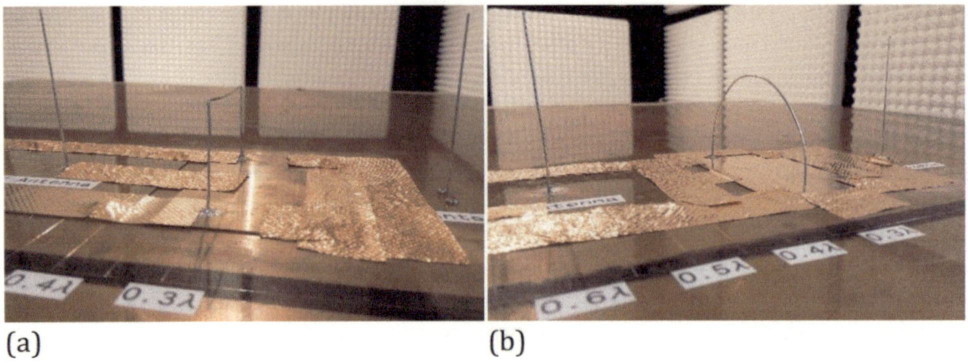

Figure 8.
Experimental setup of a wireless power transmission system integrated with a (a) rectangular and a (b) circular parasitic wire.

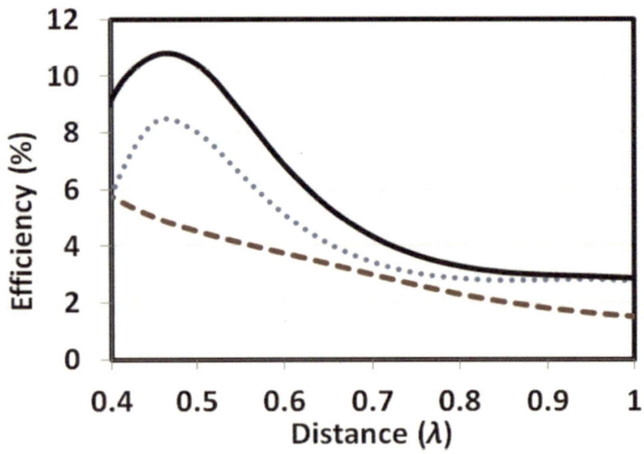

Figure 9.
The power efficiency of a conventional wireless power transmission system (dashed line) and those with a circular (dotted line) and square (solid line) parasitic wire.

antennas. Measurements were taken from the experimental setups based on the image of theory, as depicted in **Figure 8**. In the experiment, a 1 GHz signal was transferred between the 0.15-m transmitting and receiving antennas, with a square- or circle-shaped parasitic wire inserted between them. Despite having different geometries, both parasitic wires in the figures constitute an identical size of 0.3 m. The comparison between the efficiencies of the systems with and without the parasitic wires is given in **Figure 9** [7]. The efficiencies of the system with parasitic components, as seen in the figure, increase along with distance and reach their crests at approximately 0.45λ away from the transmitting antenna. The efficiencies subsequently decrease exponentially at distance beyond the crests. Even so, it can be observed that, the efficiencies of the systems with parasitic components are consistently higher than that without. This phenomenon can be attributed to the additional energy scattered from the parasitic wires. Besides the direct electromagnetic energy radiated from the transmitting antenna, the receiving antenna also collects the additional scattered energy. Unlike their conventional counterparts that suffer from path losses at the instant waves radiate from the antenna, the WPT systems with parasitic components is clearly more effective in conserving the energy carried by the waves.

4. Conclusion

In this chapter, a general overview of near- and far-field wireless power transmission (WPT) system is presented. By sparing the physical need of wire attachment, the WPT technology inculcates portability and mobility in E&E devices. Although both the near- and far-field mechanisms allow energy to be wirelessly transferred, their underlying principles differ. In the near-field mechanism, the electrical energy harvested at the load is generated from the variations of electric or magnetic fields at the intervening space between the transmitter and receiver. The far-field mechanism, on the other hand, exploits wave radiation to carry the energy. Nonetheless, the efficiency of both mechanisms deteriorates along with distance. The distance of effective power transmission can be extended if parasitic wires are integrated into the system.

Author details

Kim Ho Yeap
Universiti Tunku Abdul Rahman, Kampar, Perak, Malaysia

*Address all correspondence to: yeapkh@utar.edu.my

References

[1] Yeap KH, Hirasawa K, Nisar H. Introduction to radio astronomy and radio telescopes. In: Yeap KH, Hirasawa K, editors. Analyzing the Physics of Radio Telescopes and Radio Astronomy. Hershey PA, US: IGI Global; 2020. pp. 1-32

[2] Marincic AS. Nikola tesla and the wireless transmission of energy. IEEE Transactions on Power Apparatus and Systems. 1982;**10**:4064-4068

[3] Байрамов НВ, Титова ЮВ. Wireless electricity. Редакционнаяколлегия. 2015: 105. Available from: http://venec.ulstu.ru/lib/disk/2015/193.pdf#page=105

[4] Rozman M. Inductive Wireless Power Transmission for Automotive Applications [Ph.D. Thesis]. Manchester Metropolitan University; 2019

[5] Lee CK, Zhong WX, Hui SR. Recent progress in mid-range wireless power transfer. In: IEEE Energy Conversion Congress and Exposition. Raleigh, NC, USA: IEEE; 2012. pp. 3819-3824

[6] Cheah ARC, Yeap KH, Yeong KC, Hirasawa K. Biologically inspired wireless power transmission system: A review. In: Ponnusamy V, Zaman N, Low TJ, Amin AHM, editors. Biologically-Inspired Energy Harvesting through Wireless Sensor Technologies. Hershey PA, US: IGI Global; 2016. pp. 27-50

[7] Yeap KH, Cheah ARC, Yeong KC, Hiraguri T, Yokoyama S, Shitara I, et al. Efficiency enhancement in a wireless power transmission system using parasitic components. IETE Journal of Research. 2020;**66**:806-814

[8] Yeap KH, Hirasawa K. Introductory chapter: Electromagnetism. In: Yeap KH, Hirasawa K, editors. Electromagnetic Fields and Waves. London, UK: InTechOpen; 2019. pp. 1-10

[9] Zhang H, Lu F, Hofmann H, Liu W, Mi CC. A four-plate compact capacitive coupler design and LCL-compensated topology for capacitive power transfer in electric vehicle charging application. IEEE Transactions on Power Electronics. 2016;**31**:8541-8551

[10] Han G, Li Q, Xie K, Liu Y, Song J. Design of capacitive coupling structure for position-insensitive wireless charging. IET Power Electronics. 2020; **13**:1946-1955

[11] Cheng DK. Field and Wave Electromagnetics. 2nd ed. US: Addison-Wesley; 1989

[12] Luo B, Zhou X, Long T, Mai R, He Z. Misalignment tolerance wireless power transfer system combining inductive and capacitive coupling. IET Electric Power Applications. 2020;**14**:1925-1932

[13] Fernandes RD, Matos JN, Carvalho NB. Resonant electrical coupling: Circuit model and first experimental results. IEEE Transactions on Microwave Theory and Techniques. 2015;**63**:2983-2990

[14] Sadiku MNO. Elements of Electromagnetics. 4th ed. New York: Oxford; 2007

[15] Pawar SD, Murugavel P, Lal DM. Effect of relative humidity and sea level pressure on electrical conductivity of air over Indian Ocean. Journal of Geophysical Research. 2009;**114**:D02205

[16] Rosa RL, Livreri P, Trigona C, Donato LD, Sorbello G. Strategies and

techniques for powering wireless sensor nodes through energy harvesting and wireless power transfer. Sensors. 2019; **19**:2660

[17] Rotenberg S, Podilchak SK, Hilario Re PD, Mateo-Segura C, Goussetis G, Lee J. Efficient rectifier for wireless power transmission systems. IEEE Transactions on Microwave Theory and Techniques. 2020;**68**:1921-1932

[18] Yeap KH, Tham CY, Nisar H, Loh SH. Analysis of probes in a rectangular waveguide. Frequenz Journal of RF-Engineering and Telecommunications. 2013;**67**:145-154

[19] Boaventura A, Collado A, Carvalho NB, Georgiadis A. Optimum behavior: Wireless power transmission system design through behavioral models and efficient synthesis techniques. IEEE Microwave Magazine. 2013;**14**:26-35

[20] Kurs A, Karalis A, Moffatt R, Joannopoulos JD, Fisher P, Soljacic M. Wireless power transfer via strongly coupled magnetic resonances. Science. 2007;**317**:83-86

[21] Yeap KH, Cheah ARC, Hirasawa K, Yeong KC, Lai KC, Nisar H. Optimization of wireless power transmission systems with parasitic wires. ACES Journal. 2017;**32**:806-812

[22] Cheah ARC, Yeap KH, Hirasawa K, Yeong KC, Nisar H. Optimization of a wireless power transmission system. In: Proceedings of the International MultiConference of Engineers and Computer Scientists. Tsim Sha Tsui, Hong Kong: IAENG; 2016. pp. 590-592

Chapter 2

New Electronic Devices for Power Converters

Moufu Kong

Abstract

Power electronic devices are crucial components of power converter systems. The evolution of power devices drives the development of power converters, including improvements in performance, reliability, and power capacity. In this chapter, the author expounds the structure, working principle, and static and dynamic characteristics of the conventional PN junction diode. And the silicon carbide (SiC) Schottky barrier diode (SBD), junction barrier Schottky (JBS) diode, trench JBS (T-JBS) diode, and sidewall-enhanced trench JBS (SET-JBS) diode are also discussed and compared. Also, the structures and properties of the gallium oxide (Ga_2O_3) SBD and heterojunction diode are also summarized. Next, the author gives a detailed analysis and discussion of the silicon power metal-oxide-semiconductor field-effect transistor (MOSFET), superjunction MOSFET, and the SiC MOSFET and JFET, and the Ga_2O_3 MOSFET. Then, the device structure and operating principle, switching characteristics, and current tailing mechanism of the insulated gate bipolar transistor (IGBT) are also analyzed and summarized in detail. Finally, the energy band structure, working principle, and switching characteristic of the gallium nitride (GaN) high-electron mobility transistor (HEMT), one of the hot devices in the current market, are also described. Finally, the summary and prospect of power electronic devices are also presented in this chapter.

Keywords: electronic device, superjunction, IGBT, SiC diode, SiC MOSFET, GaN HEMT, Ga_2O_3 diode

1. Introduction

Power electronic devices are the core components of power converters and directly affect the performance and reliability of power converters. In recent years, in addition to conventional silicon-based devices, some new electronic devices have emerged, which are widely used in power conversion systems and play an important role in the performance improvement and development of power converters. This chapter explains the traditional silicon-based power diodes and power MOSFET and also describes the structures and working principles of superjunction power MOSFET devices and IGBT power devices. More importantly, with the development of the wide-bandgap semiconductor technology, SiC diodes, SiC MOSFETs, SiC JFETs, and GaN HEMTs are also widely used in various power converters and power electronic systems, so this chapter also describes the wide-bandgap power semiconductor devices. At the same time, the ultra-wide bandgap power semiconductor devices represented by gallium oxide (Ga_2O_3) have also become a research hotspot, and this chapter also explains the Ga_2O_3 power diodes and Ga_2O_3 power MOSFETs. Finally, the development trend of new electronic devices is also summarized.

2. Power diodes

2.1 Silicon PN junction diode

Silicon power diodes are the most commonly used power electronic devices, and their basic principle is the unidirectional conductivity of the PN junction diode. When a P-type doping and a N-type doping are performed on a semiconductor material, a PN junction is formed at the interface. Due to the existence of the doping concentration gradient at the interface, the holes in the P-type region diffuse to the N-type region and recombine with the majority carrier electrons in the N-type region. Similarly, the electrons in the N-type region diffuse to the side of the P-type region and recombine with the majority carrier holes of the P-type region. And a space charge region is formed at the interface, in which the N-type side has only positive charges ionized by the donors, and the P-type region has only negative charges ionized by the acceptors. Therefore, an electric field directed from the N-type region to the P-type region is formed at the space charge region. And under the action of this electric field, the minority carrier holes in the N-type region drift toward the P-type region, and the minority carrier electrons in the P-type region also drift toward the N-type region. The diffusion and drift motions of carriers will eventually reach a dynamic equilibrium, and the width of the space charge region (dcpletion region) remains constant, and the built-in electric field (E) is also maintained constant. **Figure 1** shows the equilibrium PN Junction and its space charge region.

The diode is formed when the PN junction chip is packaged and the anode (A) and cathode (K) electrodes are led out. When the PN junction (or diode) is forward biased, since the external electric field (E_V) and the built-in electric field (E) are in

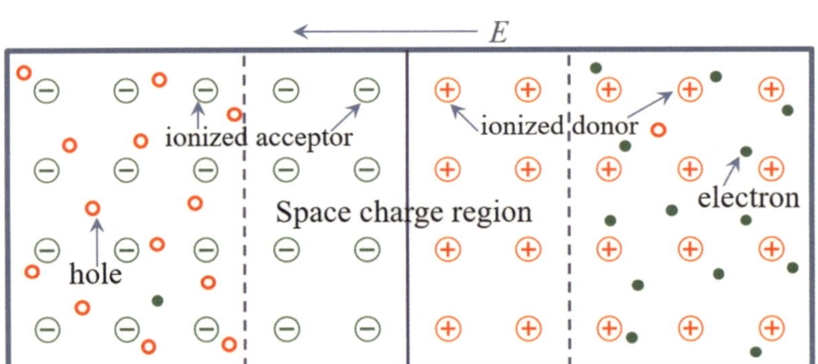

Figure 1.
Equilibrium PN junction and its space charge region.

opposite directions, the space charge region is narrowed, and the diffusion effect of the majority carriers is greatly enhanced at this time, forming a larger forward current (I_F), the PN junction is turned on, which is shown in **Figure 2**.

On the contrary, when the PN junction (or diode) is reverse biased, since the external electric field (E_V) and the built-in electric field (E) are in the same direction, the space charge region becomes wider, which greatly hinders the diffusion of majority carriers. And the PN junction is in the off state, and only a negligible reverse leakage current (I_R) flows through the diode, as shown in **Figure 3**. **Figure 3** also shows the electric field distribution of the reverse-biased diode. When the applied reverse bias voltage (V) increases, the peak electric field (E_{max}) also increases accordingly. When the E_{max} is up to the critical breakdown electric field (E_C) of the semiconductor, the diode breaks down, the applied voltage is the breakdown voltage (U_{BR}), which is also equal to the area of the electric field distribution triangle. When the diode breaks down, the reverse current of the diode will increase sharply.

The I-V characteristic curve of the PN junction diode is shown in **Figure 4**. When the forward voltage drop (V_{AK}) is higher than the turn-on (or knee) voltage drop (V_{ON}) the diode is turned ON, and the current (I_{AK}) is approximately exponential with respect to the voltage (V_{AK}). When the diode is reverse biased, its reverse leakage current (I_R) is very small and can be negligible, but when the reverse bias voltage crosses to U_{BR}, the current increases sharply, so the U_{BR} is the breakdown voltage of the diode. When designing a power converter, it is necessary to be reasonable in choosing the U_{BR} and forward current capability of the diode. The I-V characteristic of the PN diode can be described as Eq. (1).

$$\boldsymbol{I_{AK}} = \boldsymbol{I_S} \cdot \left[\mathbf{e}^{(qV_{AK}/kT)} - \mathbf{1}\right] \tag{1}$$

Where I_S is the reverse state saturation current (leakage current) I_R, T is the thermodynamic temperature, k is the Boltzmann constant, and V_{AK} is the applied bias

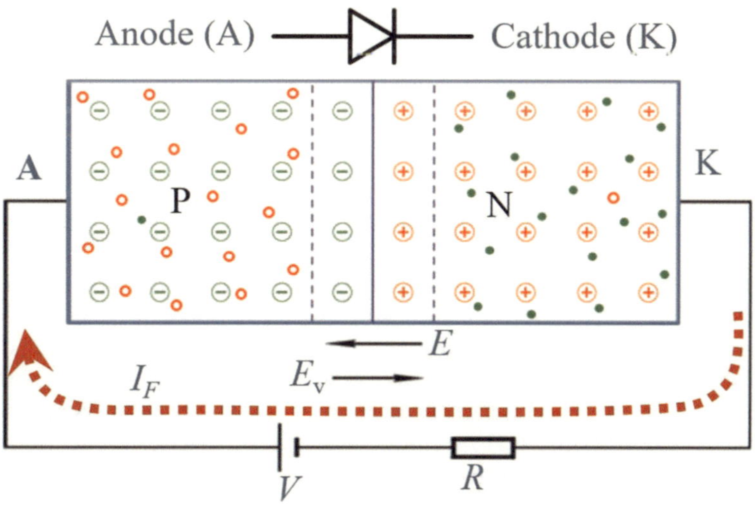

Figure 2.
PN junction or diode in the forward bias conduction state.

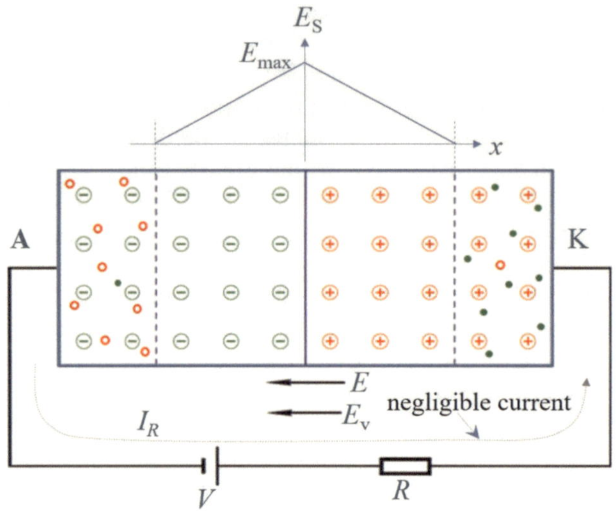

Figure 3.
PN junction diode in the reverse bias state.

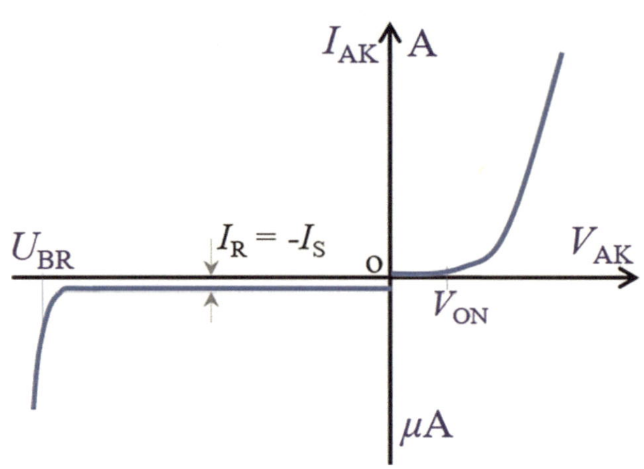

Figure 4.
The I-V characteristics of the PN junction diode.

voltage between the anode and the cathode. At room temperature, the kT/q is about 26 mV.

Due to the charge storage effect in the diode, the switching of the diode from the ON state to the OFF state requires a transient process. **Figure 5** shows the reverse recovery transient of the diode. When the power supply voltage of the diode circuit changes from the forward state to the reverse state of the diode, the current of the diode decreases from the forward on-state current (I_F) to the reverse saturation current I_R, it does not maintain at I_R immediately, but increases reversely to I_{RM}, which is because although the voltage of the external circuit has been reversed, the inside of the diode is still full of carriers, and these carriers need a process to be extracted from the body of the diode. This period of time is called the reverse recovery time (t_{rr}) of the diode, and the charge extracted during the t_{rr} time is called the reverse

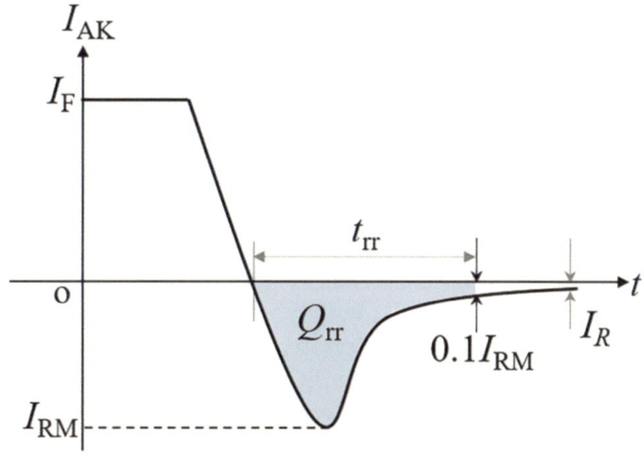

Figure 5.
The reverse recovery characteristics of the PN junction diode.

recovery charge (Q_{rr}). Generally speaking, the t_{rr} and Q_{rr} of PN junction diodes with the same rated voltage and rated current are larger than those of Schottky diodes.

2.2 SiC power diodes

The energy bandgap of 4H-silicon carbide (4H-SiC) is about 3 times that of Si (Silicon), the thermal conductivity is also 3 times that of Si, the critical breakdown electric field is about 8 to 10 times that of Si, and the saturation drift velocity of electrons is 2 times that of Si. These superior properties of SiC make it the preferred material for high-frequency, high-power, high-temperature, and radiation-resistant semiconductor devices.

Figure 6 shows the specific on-resistance ($R_{on,sp}$) of N-type drift region in 4H-SiC and Silicon at different breakdown voltages. And the $R_{on,sp}$ of the 4H-SiC drift region is about 2000 times smaller than that of the silicon devices for the same breakdown voltage [1].

Since the turn-on voltage (or knee voltage) of the SiC PN junction is as high as about 2.8 V, which is much higher than that of the SiC Schottky diode with a value of

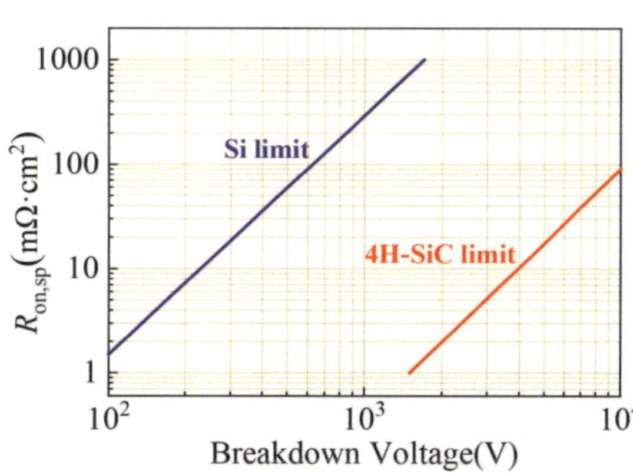

Figure 6.
Specific on-resistance of n-type drift region in 4H-SiC and silicon at at different breakdown voltages.

lower than 1 V. So, the commercial SiC diodes with a breakdown voltage of less than 4500 V are almost Schottky diodes. For the SiC Schottky diodes, the much lower drift region resistance and much higher energy bandgap compared with silicon can boost the breakdown voltage to over 3000 V with reasonable on-state voltage drop (V_F) and relatively low leakage current [1]. Although silicon-based Schottky diodes are also commonly used in power systems and power converters, they operate at low voltages (typically ≤200 V). They offer very low on-state voltage drops and losses despite high leakage current and low maximum operating temperature.

The SiC power diode structures are mainly Schottky barrier diodes (SBD) and junction barrier Schottky diodes (JBS), which are shown in **Figure 7a,b**, respectively. The main feature of the SiC SBD is the Schottky contact formed at the interface between the metal and 4H-SiC. While the SiC JBS introduces P-type regions at a certain distance in the SBD to shield the electric field at the Schottky contact interface and reduce the reverse leakage current. The energy band diagram for the metal–semiconductor (4H-SiC) contact is shown in **Figure 7c**.

In the typical normal operating state of the above two devices, the on-state current is dominated by majority carriers—electrons, and the storage effect of the minority carriers in the drift region is almost negligible. This causes the transition from the on-state to the reverse blocking state of the SiC SBD and JBS diode very fast with a much shorter reverse recovery time (t_{rr}) and a much lower reverse recovery charge (Q_{rr}) compared with those of the silicon PN diode. The high switching speed and the high current density compared with the silicon PN diodes make them suitable for high frequency, high power, and high-end applications.

For SiC Schottky power diodes, due to the low doping concentration in the N-type drift (N-drift) region to support high reverse blocking voltages, the current via thermionic emission current transport mechanism is dominant in Schottky barrier diodes. So the thermionic emission theory can be used to describe the current density J_{AK} flows across the Schottky barrier interface [2], which is shown in Eq. (2):

$$J_{AK} = A^{*}T^{2}e^{-(q\Phi_{BN}/kT)}\left[e^{(qV_{AK}/kT)} - 1\right] \tag{2}$$

where A^* is the effective Richardson constant, Φ_{BN} is the barrier height of the metal–semiconductor contact (shown in **Figure 7c**), T is the thermodynamic temperature, k is the Boltzmann constant, and V_{AK} is the applied bias voltage between the

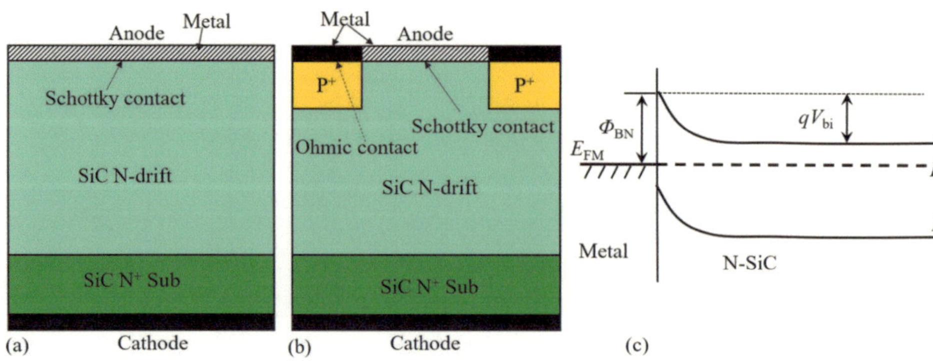

Figure 7.
(a) SiC Schottky barrier diodes (SBD), (b) SiC junction barrier Schottky diodes (JBS) and (c) the band diagram for metal–semiconductor contact.

anode and the cathode. Among them, the Richardson constant of N-type silicon carbide material is 146 $A{\cdot}cm^{-2}{\cdot}K^{-2}$.

The turn-on voltage drop (V_{ON}) of the Schottky diode is mainly determined by the barrier height (Φ_{BN}). And the Φ_{BN} mainly depends on the workfunction of metal materials, thus the on-state voltage drops of SiC Schottky diode can be selected by choosing different metal materials.

Based on the operation mechanism of SiC Schottky diode, the static I-V characteristic curves of SiC SBD and JBS diodes are shown in **Figure 8**. As can be seen from the figure, the SiC SBD exhibits a higher forward current density, but it also shows a larger leakage current and a lower breakdown voltage (U_{BR}). Although the SiC JBS diode has a lower current density, the leakage current in the blocking state is much lower and the U_{BR} is also higher than those of the SBD. This is because the P-type regions introduced in the JBS structure reduce the Schottky contact area, resulting in a certain reduction in current density, but the introduction of the P-type regions shields the electric field at the Schottky contact interface, thereby effectively reduces the reverse leakage current and improves the breakdown voltage. However, in SiC materials, the ion implantation depth of the P-type regions is relatively shallow (usually <1 μm), thus the electric field shielding effect is limited. Then, the trench JBS (T-JBS) diode has been proposed to achieve a low reverse leakage current [3, 4]. Unfortunately, the T-JBS structure introduces a severe JFET (junction field-effect transistor) effect, which greatly reduces the on-state current density. And the sidewall-enhanced JBS (SET-JBS) diode was proposed to alleviate the JFET effect and increase the Schottky contact area, resulting in an improvement in current density [5]. **Figure 9a,b** shows the device structures of the T-JBS diode and the SET-JBS diode, respectively. And the static I-V characteristic comparison result is also shown in **Figure 9c**.

2.3 Ga_2O_3 power diodes

Gallium oxide (Ga_2O_3) is a representative material of the ultra-wide bandgap semiconductor material and has attracted extensive research interest in recent years. There are five isomers of Ga_2O_3, and the beta-Ga_2O_3 (β-Ga_2O_3) is mostly used material for power devices. Due to its ultra-wideband gap (over 4 times of Si), high theoretical breakdown electric field (8MV/cm), large Baliga figure of merit (3400), and stable chemical properties, it has become an ideal choice for high-voltage and

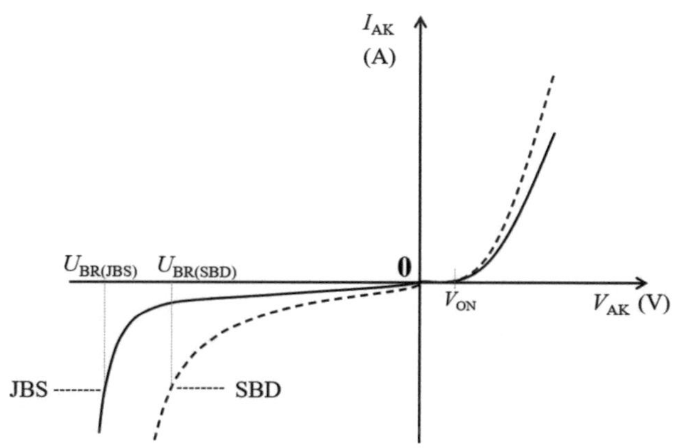

Figure 8.
The static characteristic curves of SiC SBD and JBS diodes.

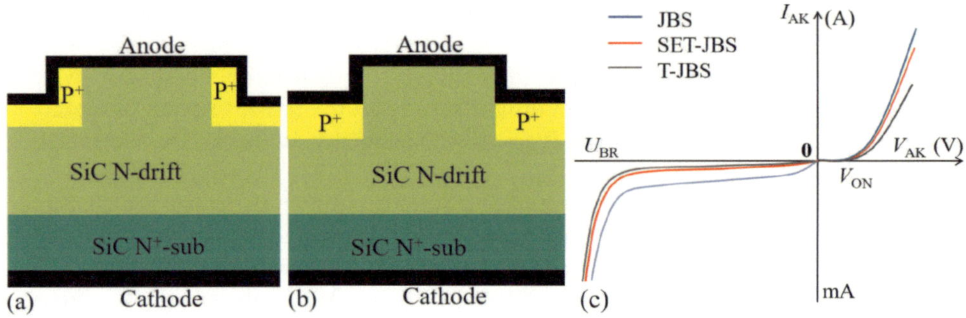

Figure 9.
(a) SiC trench JBS (T-JBS) diode, (b) SiC sidewall enhanced JBS diode (SET-JBS), and (c) static characteristic comparison of different Schottky diodes [5].

high-power rectifiers and field-effect transistors [6, 7]. Nowadays, with the improvement of crystal growth technology, large-scale Ga_2O_3 single crystals have been produced, such as pulling method, guided mode method, and floating zone melting method [8]. There is also a method of heteroepitaxial growth of gallium oxide thin films on substrates such as quartz glass, sapphire, silicon, and gallium arsenide [9–11].

Figure 10a shows the cross-sectional structure view of the Ga_2O_3 Schottky barrier diode (SBD). As shown in the figure, the top metal layer (Pt/Ti/Au) is in contact with the Ga_2O_3 N-type drift region to form a Schottky contact. And the bottom metal (Ti/Au) is in contact with a heavily doped N+ Ga_2O_3 substrate to form an ohmic contact. The operating mechanism of Ga_2O_3 SBD is similar to that of SiC SBD. However, compared with SiC materials, the effective P-type doping has not yet been achieved in Ga_2O_3 materials. Due to the lack of P-type doping in the Ga_2O_3 materials, **Figure 10b** shows a structure of a Ga_2O_3 heterojunction (HJ) PN diode [12]. The current density of the Ga_2O_3 heterojunction PN diode is higher than that of the Ga_2O_3 SBD due to the hole injection and conductance modulation effect in the N-type drift region. However, because the barrier height of the heterojunction PN junction is higher than that of Schottky contact, the turn-on voltage drop of the HJ PN diode is higher than that of the SBD, but the Ga_2O_3 heterojunction PN diode has great advantages in the field of ultra-high voltage (e.g., > 6500 V) applications with a lower voltage drop at the same current density compared with Ga_2O_3 SBD. At present, the Ga_2O_3 power heterojunction diode with a breakdown voltage exceeding 8000 V has been developed [12].

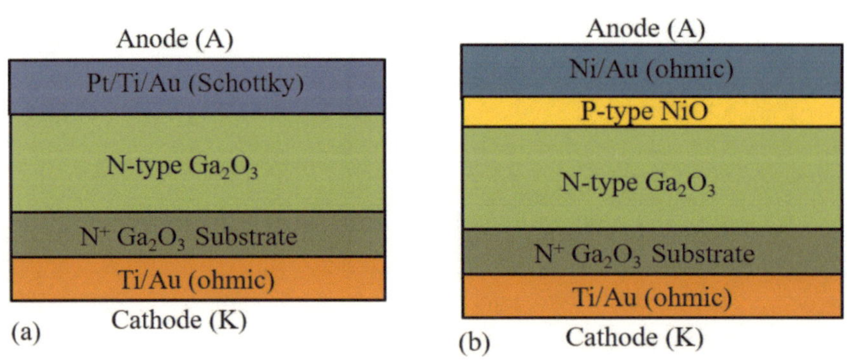

Figure 10.
Device structures of (a) Ga_2O_3 SBD, (b) Ga_2O_3 heterojunction PN diode [9].

Figure 11 shows the forward I-V characteristic curves of the Ga_2O_3 SBD and the Ga_2O_3 HJ PN diode. And it can be seen from the figure that the turn-on voltage drop of the Ga_2O_3 SBD (V_{ON1}) is lower than that of the Ga_2O_3 HJ PN diode (V_{ON2}), but the current I_{AK} of the Ga_2O_3 HJ PN diode rises faster with the voltage V_{AK}, which is mainly caused by the conductance modulation effect in the Ga_2O_3 HJ PN diode.

2.4 Dynamic characteristic comparison of the diodes

Figure 12 presents an intuitive rough comparison result of the reverse recovery characteristics of Si, SiC, and Ga_2O_3 diodes at the same rated breakdown voltage and rated current. It can be seen from the figure that the Si PN junction diode has the largest reverse recovery current, the longest reverse recovery time, and the largest

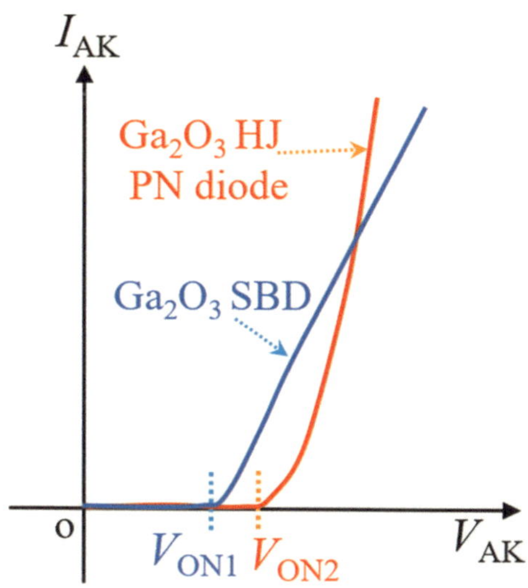

Figure 11.
The Ga_2O_3 SBD and Ga_2O_3 HJ PN diode forward I-V characteristics.

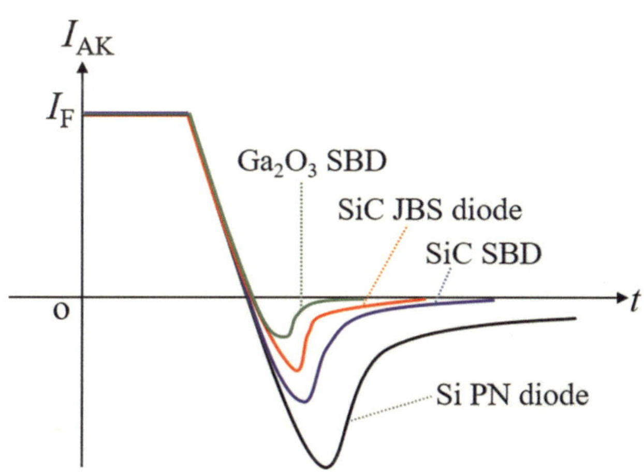

Figure 12.
Comparison of reverse recovery characteristics of Si, SiC, and Ga_2O_3 diodes at the same rated breakdown voltage and rated current.

reverse recovery charge. And the Ga_2O_3 SBD has the best reverse recovery performance. The reverse recovery performance of the SiC Schottky diodes is between that of Si and Ga_2O_3 diodes.

3. Power MOSFET and JFET

3.1 Device structure and operating mechanism

The power MOSFET (metal-oxide-semiconductor field-effect transistor) has many advantages: (a) it is a voltage-controlled device with high input impedance and low driving power consumption, (b) it is no secondary breakdown with wide safe operating area (SOA), and due to good thermal stability, the operating temperature can reach up to 200°C, which is 50°C higher than that of the bipolar transistor (BJT), (c) it is a majority carrier conduction device with the strong anti-irradiation ability and (d) it is no minority-carrier storage effect, and the switching frequency is high. Because the power MOSFET has those advantages mentioned above, it has always been a research hotspot in the industry and widely used in the power converters.

Figure 13a,b shows the device structure of the N-channel vertical power MOSFET and its electric field distribution in the blocking state, respectively. When the gate (G) to source (S) voltage (V_{GS}) is higher than the threshold voltage (V_{TH}) of the N-channel MOSFET and the drain-to-source voltage (V_{DS}) is positive, the MOSFET is turned ON, and the electrons flow from the source n + region through the channel and n-drift region to the n + drain region to form the drain current I_{DS}. Conversely, if $V_{GS} < V_{TH}$ and V_{DS} is a positive high voltage, the channel is off and the MOSFET is in a forward blocking state (voltage sustaining state). At this time, the electric field distribution inside the device is roughly as shown in **Figure 13b**, and the voltage is mainly sustained by the n-drift region. **Figure 13c,d** demonstrates the symbols of the N-channel power MOSFET and P-channel power MOSFET, respectively.

3.1.1 Static characteristic

Figure 14a,b shows the transfer characteristic and the output characteristic of the N-channel power MOSFET, respectively. And when $V_{GS} > V_{TH}$, the MOSFET is

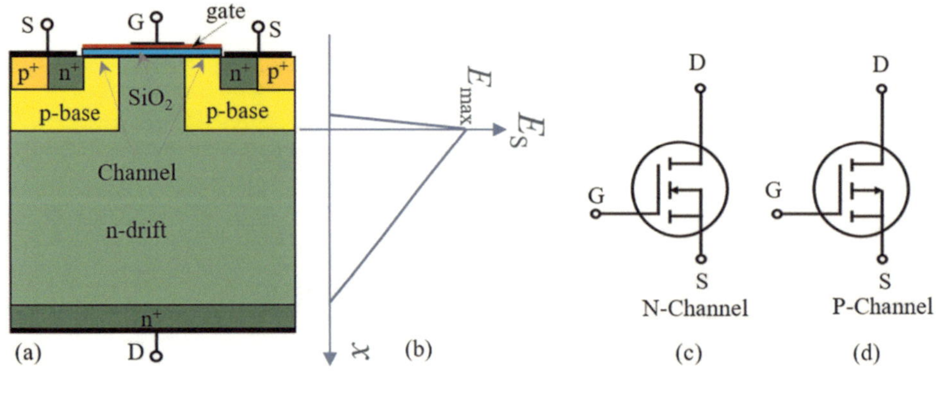

Figure 13.
(a) Device structure of the N-channel power MOSFET, (b) the electric field distribution of the power MOSFET in the blocking state; the symbols of the (c) N-channel power MOSFET and (d) P-channel power MOSFET.

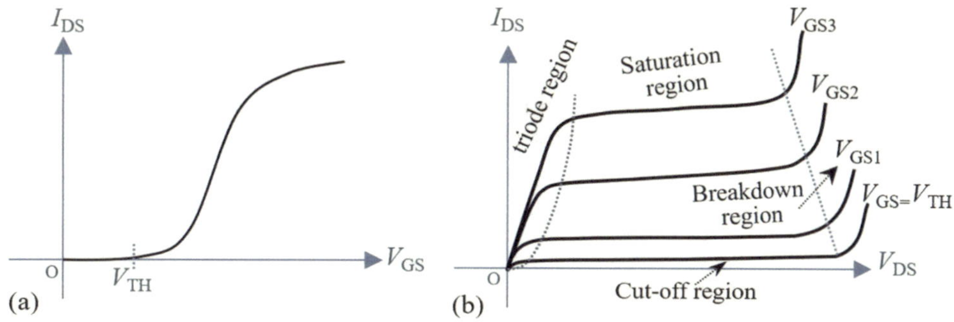

Figure 14.
(a) The transfer characteristic curve and (b) output characteristic curve of the power MOSFET.

turned ON, and the drain current I_{DS} increases as the V_{GS} increases. From the output characteristic curve, when the device is in the cut-off region, I_{DS} is almost negligible; when the device is in the triode region, the current I_{DS} increases sharply with V_{DS}. While in the saturation region, I_{DS} hardly increases with V_{DS}. And when the device is in the breakdown region, the current I_{DS} increases sharply, causing a dramatic increment in power consumption, which may cause a thermal runaway of the device. Therefore, the device should be avoided as much as possible to operate in the breakdown region.

3.1.2 Dynamic characteristic

Figure 15a,b shows the power MOSFET switching test circuit with an inductive load L and the typical characteristics of power MOSFET switching transients [13]. As the power MOSFET is a majority carrier device, there is no minority carrier storage effect, so the switching speed is fast, typically 20–50 ns. As shown in **Figure 15b**, the turning-on time of the device is $t_{sw(on)}$ ($t_{sw(on)} = t_{d1} + t_{on}$), and the turn-off time is $t_{sw(off)}$ ($t_{sw(off)} = t_{d2} + t_{off}$). Typically, most devices have a longer turn-off time than their turn-on time.

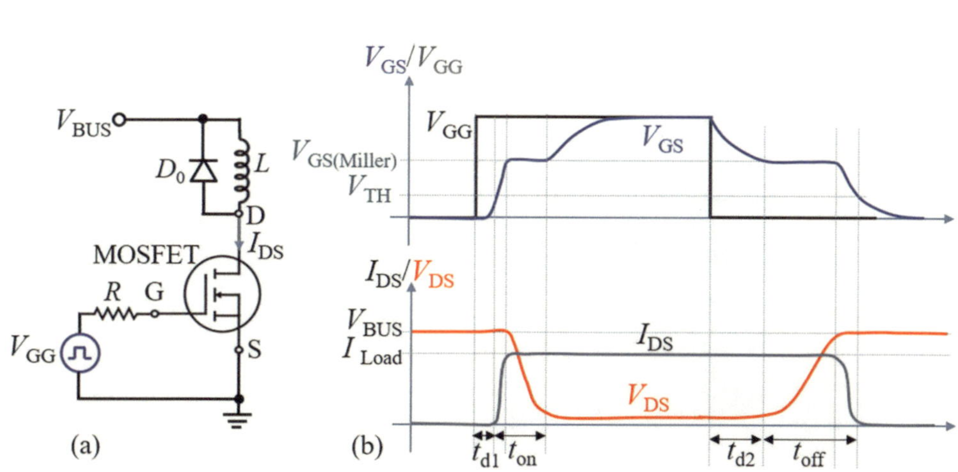

Figure 15.
(a) MOSFET switching test circuit with inductive load, (b) typical characteristics of power MOSFET switching transients [13].

3.2 Superjunction power MOSFET

The conventional power MOSFET devices have an inherent contradiction that the specific on-resistance ($R_{on,sp}$) is proportional to the 2.5th power of the breakdown voltage (BV), that is $R_{on,sp} \propto BV^{2.5}$, which is dreaded as "silicon limit" theory. This means that even with a small increase in BV, $R_{on,sp}$ will increase substantially, thereby greatly increasing the conduction loss of the device.

In 1993, Chen invented the superjunction device, which greatly improved the contradiction between $R_{on,sp}$ and BV with a much better relationship as $R_{on,sp} \propto BV^{1.32}$ [14, 15]. And the superjunction MOSFET was commercialized in 1998 and hailed as a "milestone" in the field of power electronic devices [16].

Figure 16a,b illustrate the device structure of superjunction MOSFET and its approximate electric field distribution in the drift region. Due to the introduction of the charge compensation effect of the P-type pillars in the drift region, the doping concentration of the n-drift can be greatly increased, thereby greatly reducing the $R_{on,sp.}$ of the device. At the same time, the charge compensation effect makes the total net charge of the properly designed superjunction drift region to be zero in the blocking state, so that the electric field distribution is approximately rectangular. Therefore, at the same rated voltage, the drift region thickness of superjunction MOSFETs is thinner than that of conventional power MOSFETs with a triangular electric field distribution. The much higher n-drift doping concentration and thinner n-drift region thickness greatly reduce the $R_{on,sp.}$ of the device, enabling superjunction devices to break the "silicon limit" of conventional MOSFETs.

Figure 17 shows the $R_{on,sp.}$ comparison result between the silicon conventional power MOSFET and superjunction MOSFET at different breakdown voltages [17]. As can be seen from the figure, the relationship between the $R_{on,sp.}$ and BV of the superjunction MOSFET is approximately linear. The comparison results show that the superjunction devices can be used for higher power density and higher-end applications.

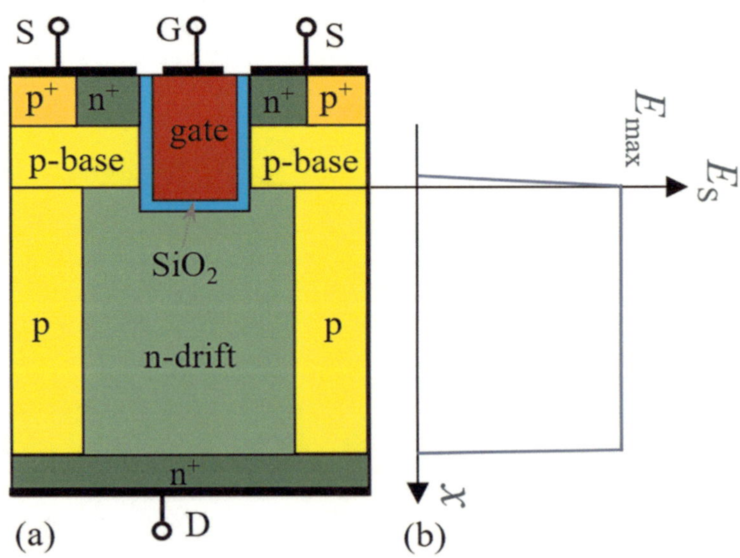

Figure 16.
(a) Structure of superjunction MOSFET and (b) the approximate electric field distribution in the drift region.

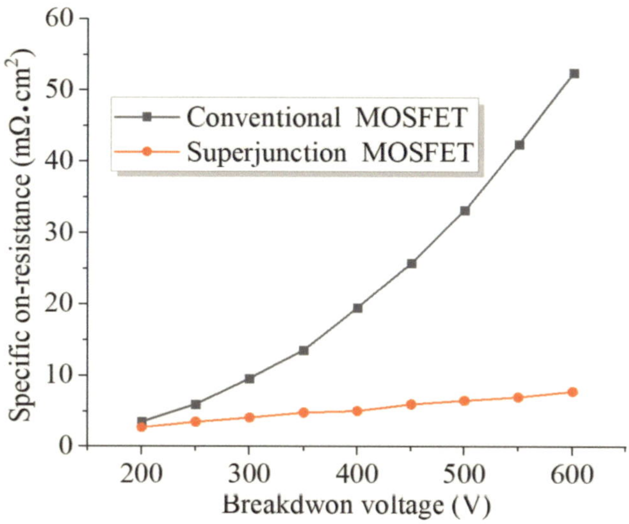

Figure 17.
$R_{on,sp}$ comparison between the silicon conventional MOSFET and Superjunction MOSFET at different breakdown voltages [17].

3.3 SiC power MOSFET

Figure 18 presents the structural comparison of Si and SiC vertical power MOSFETs at the same rated breakdown voltage. It can be seen from the figure that under the same rated breakdown voltage, the thickness of the n-drift region of the SiC MOSFET is about 1/10 of that of the silicon MOSFET, so the drift region resistance is dramatically reduced and the current density of the SiC power MOSFET is greatly improved, the conduction loss is greatly reduced, the switching speed is also improved, and the chip size is greatly reduced. At the same time, due to the larger energy band gap and higher thermal conductivity of SiC materials, the SiC MOSFETs can be operated at temperatures over 200°C. However, compared with silicon

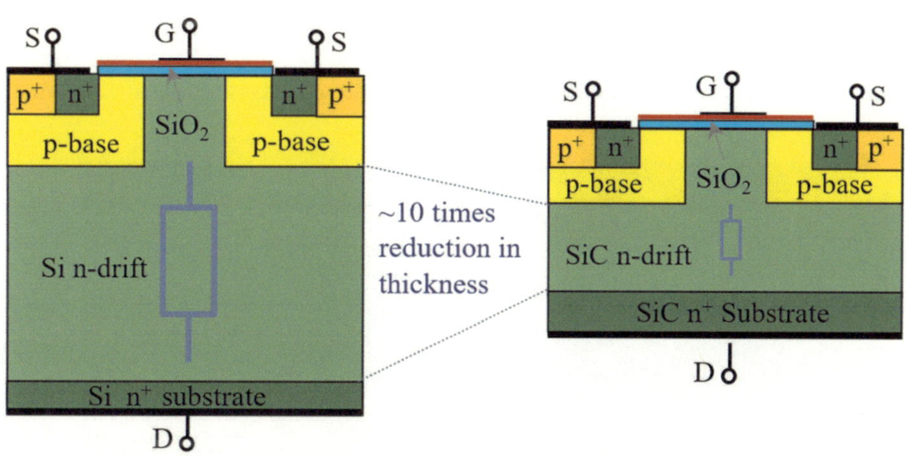

Figure 18.
Structural comparison of Si and SiC vertical power MOSFETs at the same breakdown voltage.

MOSFETs, the channel mobility of SiC MOSFETs is still very low, and its on-state resistance still has a large room for improvement. At the same time, the electric field of the gate oxide layer may be very high, which brings challenges to the reliability of the gate oxide layer.

Since the current capability of SiC devices is much larger than that of silicon devices, SiC lateral devices can also be used in power-integrated circuits to handle larger power conversions. And the SiC lateral device can easily achieve over 1200 V breakdown voltage while still obtaining a low on-resistance [18, 19]. **Figure 19** shows the cross-sectional view of a SiC lateral power MOSFET device [20]. Compared with the SiC vertical MOSFET, all electrodes of the SiC lateral power MOSFET are on the surface of the device, so that the SiC lateral MOSFET can be integrated with SiC low-voltage integrated circuits on the same chip to realize monolithic SiC power integrated circuits. The operating principle of the SiC lateral power MOSFET is almost the same as the vertical power MOSFET. The only difference between the two kinds of devices is that the current of the lateral power MOSFET flows laterally, and the electron flow path is shown by the red dotted line in the figure. It is worth mentioning that the purpose of introducing the p-top region into the lateral power MOSFET of this device is to increase the doping concentration of n-drift through the principle of charge compensation, and greatly reduce the on-resistance while optimizing the surface electric field. So the device has the advantages of high breakdown voltage and low on-resistance.

3.4 SiC JFET

Due to the extremely low channel mobility of SiC MOSFETs and the reliability issues of the SiO_2 gate oxide layer, SiC JFET (junction field-effect transistor) devices were once favored by researchers and the industry, and have also been commercialized [21, 22]. **Figure 20a,b** show the structure and the symbol of the SiC power JFET device, respectively. Different from the MOSFET, the SiC JFET controls the turn-off of the device by applying a negative voltage through the p^+ gate with respect to the source to completely deplete the N-type channel region. Generally, the SiC JFET is a normally on device, and when the gate-to-source is zero biased (V_{GS} = 0 V), the channel is not fully depleted, the device is in the on-state, and electrons flow from the n^+ source through the channel region to the n^+ drain. Since the channel mobility of SiC JFETs is much greater than that of SiC MOSFETs, the SiC JFETs have lower on-resistances. However, the gate control voltages of the two devices are different. The

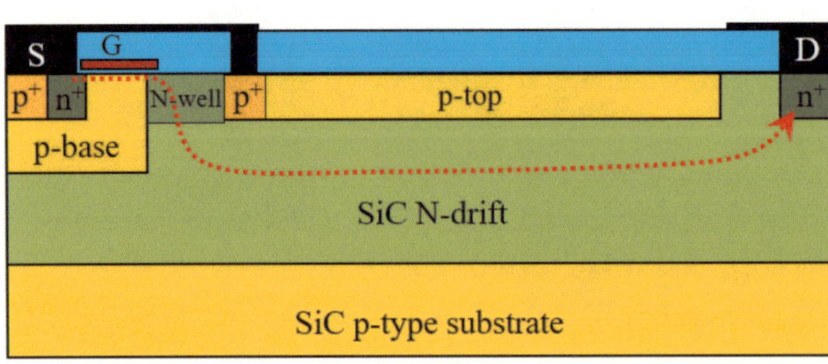

Figure 19.
Cross-sectional view of a SiC lateral power MOSFET device [18].

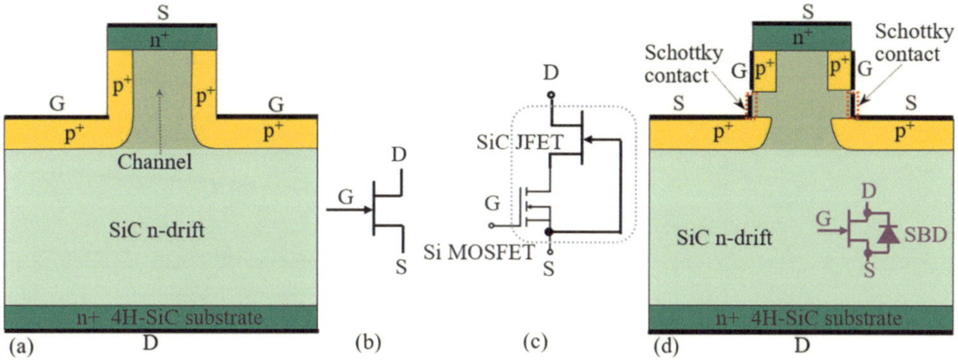

Figure 20.
(a) The cross-sectional view and (b) symbol of the SiC power JFET; (c) SiC JFET/Si MOSFET cascode configuration; (d) SiC SBD-JFET.

V_{GS} of the JFET cannot be higher than the turn-on voltage drop of the gate-source PN junction ($\sim$ 2.8 V), while the V_{GS} of the MOSFET can be as high as 20 V.

In order to take the performance advantages of SiC JFETs and make SiC JFETs as easy to control as MOSFETs at the same time, a cascode configuration consisting of a low-voltage high-current Si MOSFET and a high-voltage SiC JFET has emerged on the market and gained lots of applications, which is shown in **Figure 20c** [23]. In addition, in order to realize the self-reverse recovery of the SiC JFET device and improve its performance, a new SiC SBD-JFET has been proposed in **Figure 20d** [24].

3.5 Ga_2O_3 power MOSFET

Figure 21a shows the structure of the depletion-mode Ga_2O_3 MOSFET, which has a negative threshold voltage V_{TH} [25]. The two heavily Si-doped N$^+$ regions are connected with metals to form low-resistance ohmic contacts, respectively. The source-connected field plate above the gate can effectively reduce the surface electric field and improve the breakdown voltage of the device in the blocking state. And due to the existence of the Fe-doped semi-insulated β-Ga_2O_3 substrate, the leakage current through the substrate is obviously reduced. During forward conduction, the gate-to-source voltage V_{GS}

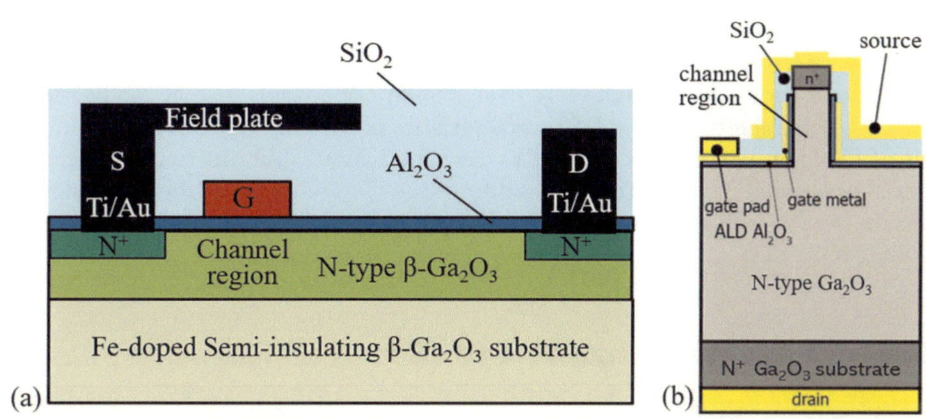

Figure 21.
(a) Depletion-mode Ga_2O_3 lateral MOSFET [25], (b) enhancement-mode Ga_2O_3 vertical transistor [26].

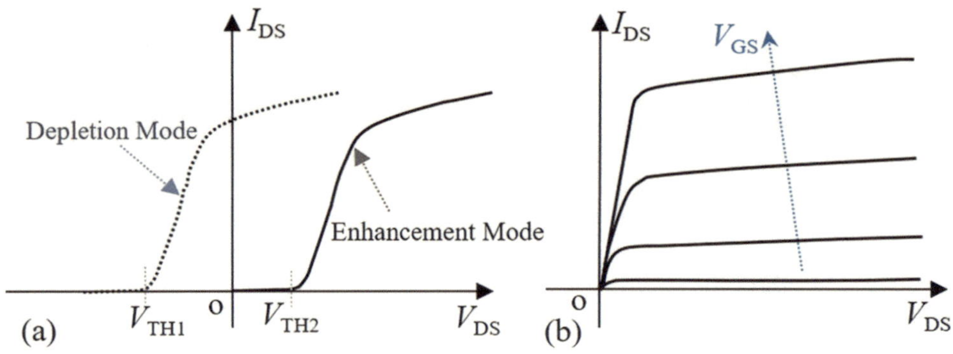

Figure 22.
(a) Transfer characteristic curve of the depletion-mode and enhancement-mode Ga_2O_3 MOSFETs (b) I-V output characteristic curve of Ga_2O_3 MOSFETs.

is higher than V_{TH}, the channel region under the gate is not fully depleted, and a positive drain voltage relative to the source V_{DS} is applied, the electrons flow from the N^+ source to the drain along the channel region and the N-type β-Ga_2O_3 drift region to form the on-state drain current I_{DS}. When the gate-to-source voltage $V_{GS} < V_{TH}$, the channel region is fully depleted, and the device is changed to the blocking state with no electron flowing from the source to drain. And the maximum blocking voltage (breakdown voltage) is mainly determined by the distance from the gate to the drain, the doping concentration of N-type β-Ga_2O_3, and the parameters of the field plate. **Figure 21b** shows a vertical enhancement-mode Ga_2O_3 MOSFET with a breakdown voltage over 1 kV [26].

The transfer characteristic curve of the Ga_2O_3 MOSFET is shown in **Figure 22a**. It can be seen from the figure that the V_{TH} of the depletion-mode device is negative, and the V_{TH} of the enhancement-mode device is positive. Since there is no effective P-type doping in Ga_2O_3, therefore, the common devices are almost in depletion mode. The main way of realizing enhancement-mode Ga_2O_3 MOSFET is to make the channel region to be very thin (such as the recessed gate structure [27]) or to be very narrow (as shown in **Figure 21b**), so that when the V_{GS} is zero biased, the channel region can also be completely depleted. Although the realization of enhancement-mode devices increases the complexity of the fabrication process, the enhancement-mode devices are easier to be controlled from the application point of view. **Figure 22b** plots the output I-V curves of the Ga_2O_3 MOSFET and both the depletion-mode and enhancement-mode Ga_2O_3 MOSFETs have similar I-V curves. Also, it can be seen that the transfer characteristics and I-V characteristics of the Ga_2O_3 MOSFET are similar to those of the Si and SiC power MOSFETs described above, but due to the different material parameters, the current capability of the Ga_2O_3 MOSFET device is higher. In addition, the switching characteristics of Ga_2O_3 MOSFETs are similar to those of Si and SiC MOSFETs.

4. Insulated gate bipolar transistor (IGBT)

Insulated gate bipolar transistor (IGBT) is a composite fully controlled voltage-driven power electronic device composed of BJT (bipolar junction transistor) and MOSFET, which has both the high input impedance and the low on-state voltage drop.

It was once hailed as an almost ideal switching device in the field of power electronics, except for its higher turn-off loss and longer turn-off time compared with those of the power MOSFETs.

Figure 23a,b demonstrate cross-section view of the IGBT and the equivalent circuit of internal structure, respectively [28]. From the perspective of device structure, the n^+ region connected to the emitter (E) electrode, p-base, gate, and n-drift region constitute an N-channel MOSFET (N-MOSFET). At the same time, the p^+/p-base, n-drift/n-buffer regions, and the p^+ collector (C) constitute a PNP BJT. Therefore, from the perspective of the internal device structure, the IGBT can be regarded as a combination of an N-MOSFET and a PNP BJT. Thus, the IGBT has the advantages of the high input resistance of N-MOSFET and the large current density of BJT. It is worth noting that many researchers also call the electrode collector (C) on the back of the device as anode (A), and the electrode emitter (E) on the surface as cathode (K). **Figure 23c** also shows the symbol of the IGBT.

In the blocking state, the gate voltage with respect to the emitter (V_{GE}) is zero biased or negatively biased. At this time, the N-MOSFET controlled by the gate is in the off-state, and the IGBT is also in the off-state. And the applied positive voltage between the collector electrode and the emitter electrode (V_{CE}) is sustained by the P-base/N-drift junction. Since the doping concentration of the p-base region is much higher than that of the n-drift region, so the breakdown voltage of the IGBT is mainly determined by the thickness and doping concentration of the n-drift region.

In the on-state, V_{GE} is applied to a positive voltage (usually +15 V). And an inversion layer electron channel connecting the n^+ region and the n-drift region is formed on the surface of the p-base region under the gate. The electrons flow from the n^+ region through the channel into the n-drift region and finally into the p^+ collector region. The electron current acts as the base drive current of the PNP transistor, which facilitates the injection of holes from the p^+ collector region into the n-buffer and n-drift regions, and finally into the emitter electrode.

Figure 24a,b illustrate the transfer characteristic and output characteristic curves of the IGBT, respectively. As can be seen from the figure, the transfer characteristic curve of the IGBT is similar to that of power MOSFET. However, the forward output characteristic of the IGBTs is slightly different from that of MOSFETs. The main

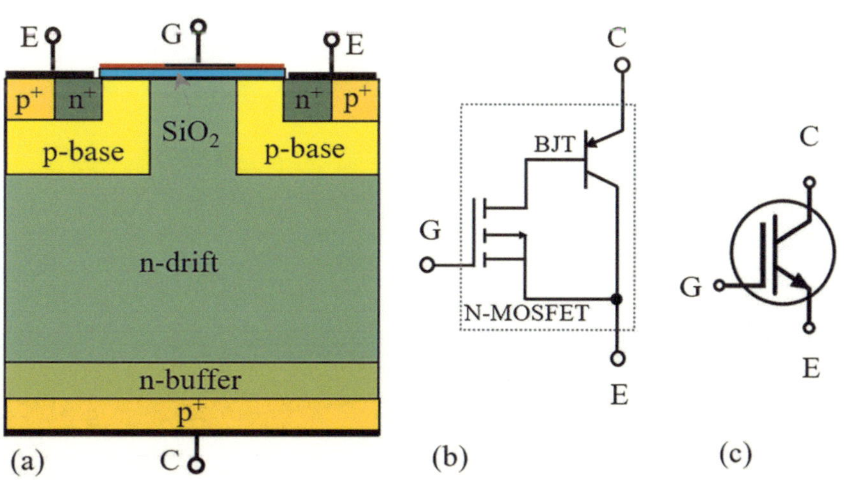

Figure 23.
(a) Cross-section view of the IGBT, (b) equivalent circuit of internal structure, (c) the symbol of the IGBT.

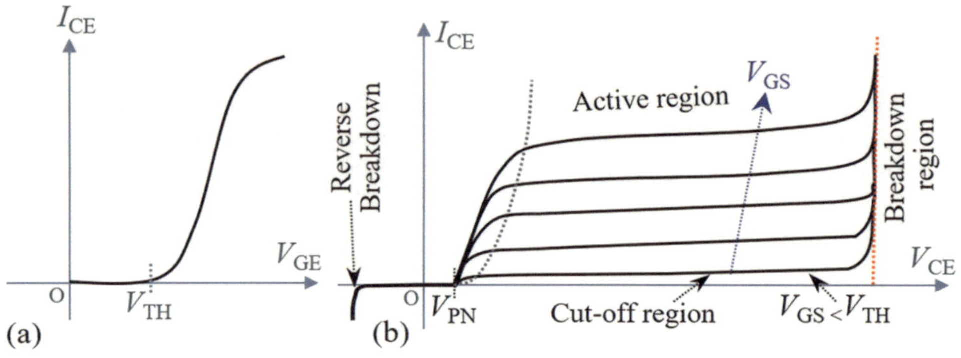

Figure 24.
(a) Transfer characteristic and (b) output characteristic curves of the IGBT.

difference is that, in addition to $V_{GE} > V_{TH}$, the forward conduction of IGBTs requires V_{CE} to be higher than the turn-on voltage V_{PN} of the p^+/n-buffer PN junction. After the turning-on of the device, a large number of holes are injected into the n-drift region from the p^+ collector, resulting in a conductance modulation effect, which greatly increases the current density and reduces the on-state voltage drop (V_{ON}) of the device. **Figure 24b** also reveals the reverse breakdown characteristic of the IGBT, the reverse breakdown voltage is mainly determined by the breakdown voltage of the p^+/n-buffer junction, usually because the doping concentration of the n-buffer region is much higher than the n-drift region, the reverse breakdown voltage of the IGBT is very low. However, for an IGBT without an n-buffer region, the reverse breakdown voltage may also be close to the forward breakdown voltage.

The turning-on characteristic of the IGBT is similar to that of the power MOSFET, but its turning-off characteristic is different from that of the power MOSFET. **Figure 25** shows the typical switching-off characteristic of the IGBT [29]. Compared with the switching-off transient of the power MOSFET, the turning-off transient of the IGBT is much longer and a long tail current is appeared in the turning-off process [30]. The reason is that after the channel of the IGBT is turned off, a large number of nonequilibrium electrons in the drift region flow out to the p^+ collector region under the action of the electric field. During this transient, the bottom p^+/n-buffer PN junction is still in the forward biased state, and holes are continuously injected into the

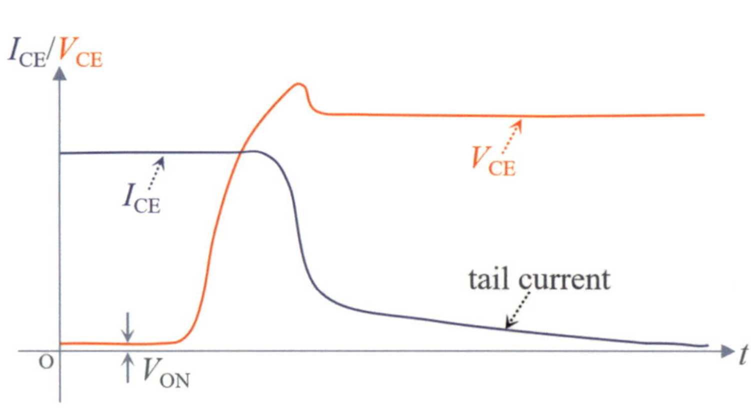

Figure 25.
Typical switching off characteristic of the IGBT [29].

n-drift region until almost all the electrons are extracted from the n-drift region. Thus, the holes injection during the turn-off process is the main reason for the current tailing. Although there are some design optimizations and structural improvements to increase the turn-off speed of IGBTs, most of them come at the expense of forward voltage drop (V_{ON}) [31]. And the improvements in the trade-off relationship between the turn-on voltage drop and turn-off loss of the IGBT are still being pursued [32]. Nevertheless, the switching speed of IGBT is far inferior to that of MOSFET, which limits the high-frequency application of the IGBT.

5. GaN high electron mobility transistor (HEMT)

In 1992, the first Gallium nitride high electron mobility transistor (GaN HEMT) was developed by using metal organic chemical vapor deposition (MOCVD) [33]. Subsequently, GaN devices have attracted great attention and research. The GaN HEMT has the advantages of high current density, high breakdown voltage, high operating frequency, high reliability, and low switching loss. GaN HEMTs have great potential for application in high frequency, high efficiency, and high power density power electronic systems. Currently, the GaN HEMTs are commercialized with rated voltage up to 650 V and are widely used in power converters, power adapters, on-board charging, data centers, and other applications [34].

For nitride semiconductors, the thermodynamic stable phase is a hexagonal symmetric wurtzite structure, while the thermodynamic metastable phase is a cubic symmetric sphalerite structure [35]. The wurtzite GaN crystal structure does not show symmetry along the C-axis, besides the sum of the vector P of the polarization intensity of Ga-N covalent bond is not zero. There should be a deviation between positive and negative ions, so a strong spontaneous polarization effect generates inside the GaN. As for AlGaN/GaN heterojunctions, the lattice constants of the two materials do not match, leading to the existence of stress forces between atoms near the contact surface of the two materials. Under the action of this stress force, the lattice asymmetry is enhanced, meanwhile the lattice deformation deviates the center of positive and negative charges in the lattice, resulting in a strong piezoelectric polarization effect.

Figure 26a,b show the energy band diagram and the structure of the enhancement-mode GaN HEMT, respectively [36, 37]. The heterojunction energy is discontinuous due to the strong total polarized induced electric field generated by the

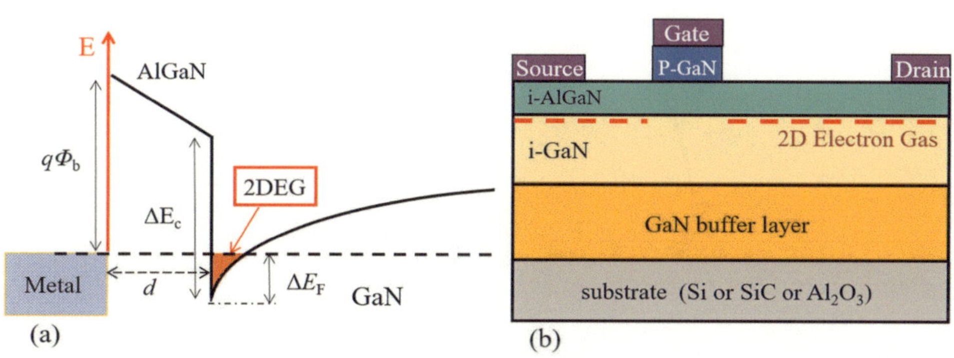

Figure 26.
(a) GaN HEMT basic energy band diagram, (b) enhancement-mode GaN HEMT basic configuration [36, 37].

addition of piezoelectric polarization and spontaneous polarization in the heterostructure, as well as the conduction band shift ΔE_c at the heterojunction interface. The energy band bends in the GaN layer to form a triangular electron potential well, which captures electrons ionized by donor impurities and then forms a large number of two-dimensional electron gas (2DEG). Due to the existence of a high potential barrier on the side of AlGaN, it is hard for electrons to cross the potential well, therefore, electrons are restricted to move laterally in the thin layer of the interface, instead of moving perpendicular to the interface. Different from the channel electrons in traditional MOSFET, 2DEG accumulates on one side of the intrinsic potential well layer, realizing the separation of carriers and the Coulomb scattering center. There is almost no electron impurity scattering in the potential well, which indicates that the 2DEG has a very high electron mobility. At the same time, the concentration of 2DEG obtained under the unintentional doping of GaN is quite considerable, thereby the GaN HEMT devices have high current density and unique application value.

1. Static characteristic

Figure 27a,b show the transfer characteristics and output characteristics of the enhancement-mode GaN HEMT, respectively. The threshold voltage (V_{TH}) of the GaN HEMT is the gate-to-source voltage (V_{GS}) corresponding to the device from off state to on state, that is, the V_{TH} is the voltage applied to the gate when the AlGaN and GaN interface forms the 2DEG. Currently, the threshold voltage (V_{TH}) of the typical commercial AlGaN/GaN HEMT is positive—enhancement mode HEMT.

It is found that the 2DEG characteristics are very similar to the channel electrons of MOSFET, so the output characteristics of the enhanced GaN HEMT are very close to the characteristics of n-channel MOSFET. By varying the applied drain-source voltage when the device is turned on, GaN HEMT can be operated in the linear (unsaturated) and saturation regions. However, in the GaN HEMT, when V_{DS} is high, the drain current I_{DS} decreases with the increase of V_{DS}, and the current collapse effect occurs. The main reason is that under the large V_{DS}, a high electric field is generated between the gate and drain, and the channel hot electrons are excited to tunnel to the surface of AlGaN, and are trapped by the

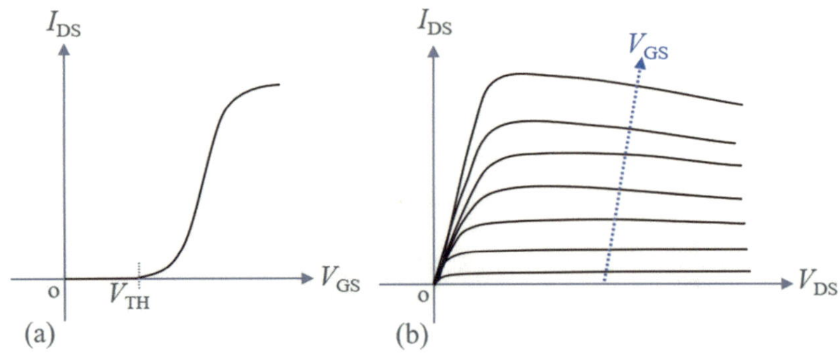

Figure 27.
(a) Enhancement mode GaN HEMT transfer characteristics curve, (b) GaN HEMT DC output characteristics curve.

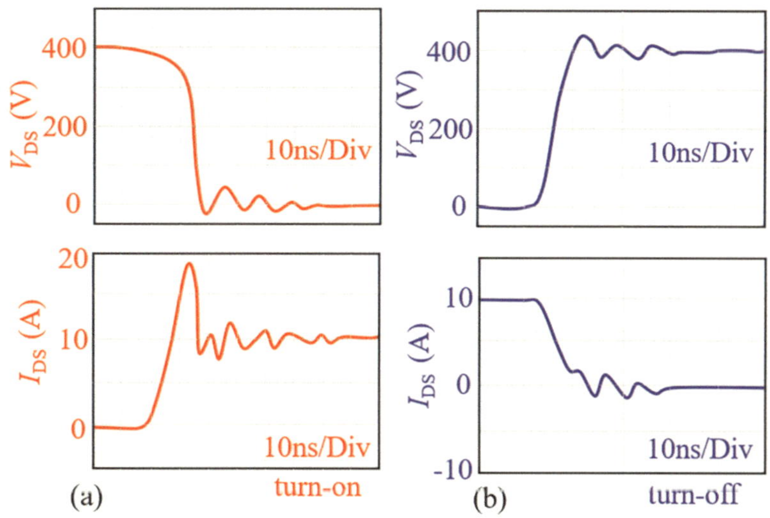

Figure 28.
Switching characteristics of the enhancement-mode GaN HEMT: (a) turn-on, (b) turn-off (redrawn from figure 1 in ref. [39]).

surface states between the gate and drain, forming a virtual gate and a current collapse phenomenon [38].

2. Dynamic characteristics

High concentration of 2DEG with high mobility exists in the channel layer of the GaN HEMT, which on the one hand enables the device to form a maximum forward current, and on the other hand enables it to operate at high frequency and high power. Compared with traditional Si-based and SiC MOSFETs, the GaN HEMT has a faster switching time and smaller switching loss. **Figure 28** shows the typical switching waveforms of an enhancement-mode GaN HEMT in a double-pulse test circuit [39]. As can be seen from the figure, the turn-on and turn-off times of the GaN HEMT are very short, both around 10 ns. However, it is worth noting that different GaN HEMT devices and applications in different circuits have slightly different switching times.

6. Prospects for new electronic devices

For power electronic devices, we always pursue higher breakdown voltage, lower loss, higher reliability and thermal stability, and low cost. In recent years, with the increasing demand for electronic devices in power electronic systems and power converters, new electronic devices represented by SiC devices and GaN devices have also achieved rapid development. However, there is still much room for improvement in the performance of these devices, and these new devices will continue to achieve breakthroughs in performance and cost reductions in the future. At the same time, besides the Ga_2O_3 devices, new electronic devices based on ultra-wide bandgap materials (such as diamond, BN, and AlN) will also emerge one after another and will be gradually applied in the market for high voltage, high-power, and high-end

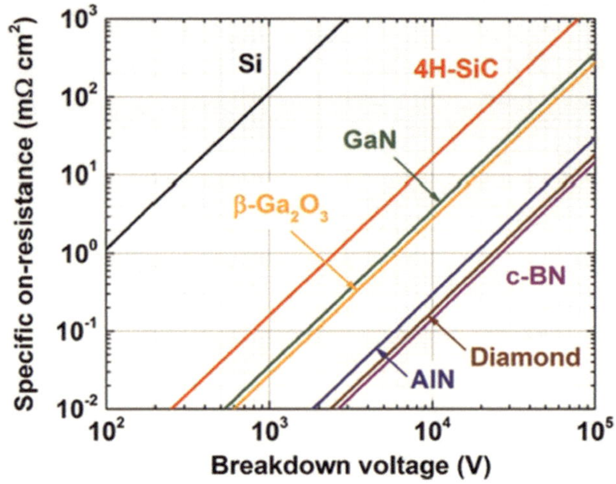

Figure 29.
Relationship between specific on-resistance and breakdown voltage of power electronic devices based on various materials [40].

applications. **Figure 29** shows the relationship between specific on-resistance and breakdown voltage of power electronic devices based on various materials [40], which also shows the development trend of power device materials from another perspective.

7. Conclusions

The emergence and development of new power electronic devices are critical to the development of power converters and power electronic systems. Understanding how electronic devices work is important for better design of power converters. This chapter describes in detail the power electronic devices commonly used in power converters. Starting from the structure and working principle of PN junction, this chapter describes the structure and main properties of SiC and Ga_2O_3 power diodes. And the structure and characteristics of power MOSFET, superjunction MOSFET, SiC MOSFET, SiC JFET, and Ga_2O_3 MOSFET are described. Then, the structure, principle, and characteristics of IGBT, an extremely important bipolar device in modern power electronic systems, are described. Finally, the structure, working principle and related characteristics of the emerging GaN HEMT devices are also described in detail. And looking ahead, new power electronic devices, such as diamond diodes and diamond MOSFETs, will continue to appear and develop to meet the more stringent requirements of power converters for lower loss, higher breakdown voltage, higher power density, higher switching frequency, and reliability.

New Electronic Devices for Power Converters
DOI: http://dx.doi.org/10.5772/intechopen.108467

Author details

Moufu Kong
State Key Laboratory of Electronic Thin Films and Integrated Devices, University of Electronic Science and Technology of China, Chengdu, China

*Address all correspondence to: kmf@uestc.edu.cn

References

[1] Baliga BJ. Wide Bandgap Semiconductor Power Devices: Materials, Physics, Design, and Applications. Cambridge, United States: Woodhead Publishing; 2018

[2] Sze SM, Li Y, Ng KK. Physics of Semiconductor Devices. 4th ed. New Jersey: Wiley; 2021

[3] Tang YD, Dong SX, Bai Y, et al. Mechanisms and characteristics of large-area high-current-density 4H-SiC trench junction barrier Schottky diodes. Materials Science Forum. 2019;**963**: 562-566. DOI: 10.4028/www.scientific.net/msf. 963.562

[4] Dou W, Song Q, Yuan H, et al. Design and fabrication of high performance 4H-SiC TJBS diodes. Journal of Crystal Growth. 2020; **2019**(533):125421. DOI: 10.1016/j.jcrysgro.2019.125421

[5] Kong M, Chen Z, Gao J, et al. A SiC sidewall enhanced trench JBS diode with improved forward performance. Semiconductor Science and Technology. 2022;**37**:075008. DOI: 10.1088/1361-6641/ac668a

[6] Murakami H, Nomura K, Goto K, et al. Homoepitaxial growth of β- Ga_2O_3 layers by halide vapor phase epitaxy. Applied Physics Express. 2015;**8**(1): 015503.1-015503.4. DOI: 10.7567/APEX.8.015503

[7] Higashiwaki M, Sasaki K, Kuramata A, et al. Gallium oxide (Ga_2O_3) metal-semiconductor field-effect transistors on single-crystal β-Ga_2O_3 (010) substrates. Applied Physics Letters. 2012;**100**(1):013504-013504-3. DOI: 10.1063/1.3674287

[8] Xia HF, Sai C, Cui Q, et al. Growth and fundamentals of bulk β-Ga_2O_3 single crystals. Journal of Semiconductors. 2019;**40**(1):011801. DOI: 10.1088/1674-4926/40/1/011801

[9] Rafique S, Han L, Neal AT, et al. Towards high-mobility Heteroepitaxial β-Ga_2O_3 on sapphire – dependence on the substrate off-Axis angle. Physica Status Solidi (a). 2018;**215**(2):1700467. DOI: 10.1002/pssa.201700467

[10] Gu K, Zhang Z, Tang K, et al. Effect of a seed layer on microstructure and electrical properties of Ga_2O_3 films on variously oriented Si substrates. Vacuum. 2022;**195**:110671. DOI: doi.org/10.1016/j.vacuum.2021.110671

[11] Chen W, Jiao T, Li Z, et al. Preparation of β-Ga_2O_3 nanostructured films by thermal oxidation of GaAs substrate. Ceramics International. 2022; **48**(4):5698-5703. DOI: 10.1016/j.ceramint.2021.11.115

[12] Zhang J, Dong P, Dang K, et al. Ultra-wide bandgap semiconductor Ga_2O_3 power diodes. Nature Communications. 2022;**13**(1):1-8. DOI: doi.org/10.1038/s41467-022-31664-y

[13] Morandin M. Electric Drives with Permanent Magnet Synchronous Machines Connected to Internal Combustion Engines [Thesis]. Padova, Italy: University of Padua. 2013

[14] Chen XB. Semiconductor Power Devices With Alternating Conductivity Type High-Voltage Breakdown Regions. China: University of Electronic Science and Technology of China; 1993

[15] Chen XB, Mawby PA, Board K, et al. Theory of a novel voltage-sustaining layer for power devices. Microelectronics Journal. 1998;**29**(12):

1005-1011. DOI: 10.1016/S00 26-2692 (98)00065-2

[16] Zhang B, Zhang W, Qiao M, et al. Concept and design of super junction devices. Journal of Semiconductors. 2018;**39**(2):021001. DOI: 10.1088/1674-4926/39/2/021001

[17] Choi W, Son D, Hallenberger M. New generation super-junction MOSFET for lower switching noise and reliable operation by controlled dv/dt and di/dt switching behavior. In: PCIM Europe. Nuremberg, Germany: VDE Verlag GmbH; 2013

[18] Kong M, Hu Z, Gao J, et al. A 1200-V-class ultra-low specific on-resistance SiC lateral MOSFET with double trench gate and VLD technique. IEEE Journal of the Electron Devices Society. 2021;**10**:83-88. DOI: 10.1109/JEDS.2021.3136341

[19] Lee WS, Lin CW, Yang MH, et al. Demonstration of 3500-V 4H-SiC lateral MOSFETs. IEEE Electron Device Letters. 2011;**32**(3):360-362. DOI: 10.1109/led.2010.210041

[20] Kong M, Duan Y, Gao J, et al. A novel optimum variation lateral doping SiC lateral double-diffused metal oxide semiconductor with improved performance. Semiconductor Science and Technology. 2022;**37**: 105022. DOI: 10.1088/1361-6641/ac88f0

[21] Kong M, Hou Y, Yi B, et al. A high speed high voltage normally-off SiC vertical JFET power device. In: 2019 IEEE International Conference on Electron Devices and Solid-State Circuits (EDSSC); 12–14 June 2019; Xi'an. New York: IEEE; 2019. pp. 1-3

[22] Kamal A, Na R, Zhou Y, et al. Ultrafast sub-mΩ battery switching module using SiC JFETs for hybrid electric aircraft propulsion applications. In: IEEE Applied Power Electronics Conference and Exposition (APEC); 14–17 June 2021. New York: IEEE; 2021. pp. 829-834

[23] Zhao C, Wang L, Yang X, et al. Comparative investigation on paralleling suitability for SiC MOSFETs and SiC/Si cascode devices. IEEE Transactions on Industrial Electronics. 2021;**69**(4): 3503-3514. DOI: 10.1109/TIE.2021.3070519

[24] Kong M, Guo J, Gao., et al. A high-performance 4H-SiC JFET with reverse recovery capability and low switching loss. IEEE Transactions on Electron Devices. 2021;68(10): 5022-5028. DOI: 10.1109/TED.2021.3101749

[25] Mun JK, Cho K, Chang W, et al. Editors' choice—2.32 kV breakdown voltage lateral β-Ga_2O_3 MOSFETs with source-connected field plate. ECS Journal of Solid State Science and Technology. 2019;**8**(7):Q3079-Q3082. DOI: 10.1149/2.0151907jss

[26] Hu Z, Nomoto K, Li W, et al. Enhancement-mode Ga_2O_3 vertical transistors with breakdown voltage> 1 kV. IEEE Electron Device Letters. 2018; **39**(6):869-872. DOI: 10.1109/LED.2018.2830184

[27] Chabak KD, McCandless JP, Moser NA, et al. Recessed-gate enhancement-mode β-Ga_2O_3 MOSFETs. IEEE Electron Device Letters. 2017;**39**(1): 67-70. DOI: 10.1109/LED.2017.27798679

[28] Baliga BJ. The IGBT Device: Physics, Design and Applications of the Insulated Gate Bipolar Transistor. William Andrew: Elsevier; 2015. pp. 3-5

[29] Alvarez R, Filsecker F, Bernet S. Characterization of a new 4.5 kV press

pack SPT+ IGBT for medium voltage converters. In: 2009 IEEE Energy Conversion Congress and Exposition; 20–24 September 2009; San Jose. New York: IEEE; 2009. pp. 3954-3962

[30] Lauritzen PO, Andersen GK, Helsper M. A Basic IGBT Model with Easy Parameter Extraction, 2001 IEEE 32nd Annual Power Electronics Specialists Conference, 17–21 June 2001, Vancouver, BC, Canada. New York: IEEE; 2022. pp. 2160-2165

[31] Iwamuro N, Laska T. IGBT history, state-of-the-art, and future prospects. IEEE Transactions on Electron Devices. 2017;**64**(3):741-752. DOI: 10.1109/TED.2017.2654599

[32] Hu H, Kong M, Yi B, et al. A novel double-RESURF SOI-LIGBT with improved Von-Eoff tradeoff and low saturation current. IEEE Transactions on Electron Devices. 2020;**67**(3): 1059-1065. DOI: 10.1109/TED.2020.2964944

[33] Khan MA, Kuznia JN, Van Hove JM, et al. Observation of a two-dimensional electron gas in low pressure metalorganic chemical vapor deposited GaN-$Al_xGa_{1-x}N$ heterojunctions. Applied Physics Letters. 1992;**60**(24): 3027-3029. DOI: 10.1063/1.10679

[34] Pushpakaran BN, Subburaj AS, Bayne SB. Commercial GaN-based power electronic systems: A review. Journal of Electronic Materials. 2020; **49**(11):6247-6262. DOI: 10.1007/s11664-020-08397-z

[35] Juza R, Hanh H. Über die Nitride der Metalle der ersten Nebengruppen des periodischen Systems. Metallamide und Metallnitride. Zeitschrift fur anorganische und allgemeine chemie. 1940;**244**(2):133-148. DOI: 10.1002/zaac.19402440205

[36] Badawi N, Hilt O, Bahat-Treidel E, et al. Investigation of the dynamic on-state resistance of 600 V normally-off and normally-on GaN HEMTs. IEEE Transactions on Industry Applications. 2016;**52**(6):4955-4964. DOI: 10.1109/TIA.2016.2585564

[37] Greco G, Iucolano F, Roccaforte F. Review of technology for normally-off HEMTs with p-GaN gate. Materials Science in Semiconductor Processing. 2018;**78**:96-106. DOI: 10.1016/j.mssp.2017.09.027

[38] Ghosh S, Das S, Dinara SM, et al. Off-state leakage and current collapse in AlGaN/GaN HEMTs: A virtual gate induced by dislocations. IEEE Transactions on Electron Devices. 2018; **65**(4):1333-1339. DOI: 10.1109/TED.2018.2808334

[39] Huang X, Liu T, Li B, et al. Evaluation and applications of 600V/650V enhancement-mode GaN devices. In: 2015 IEEE 3rd Workshop on Wide Bandgap Power Devices and Applications (WiPDA); 02–04 November 2015; Blacksburg. New York: IEEE; 2015. pp. 113-118

[40] Ray-Hua Horng. Semiconductor Ga_2O_3 Technology [Internet]. 2022. Available from: https://www.eetasia.com/the-principles-advantages-and-industry-prospects-of-fourth-generation-semiconductor-ga2o3-technology/

Chapter 3

Automatic Current Sharing Mechanism in Two-Phase Series Capacitor Buck DC-DC Converter (2-pscB)

Salahaldein A. Rmila

Abstract

In this chapter, we introduce the concept of the Inherited Automatic Current Sharing Mechanism (ACSM) in a two-phase series capacitor buck topology (2-pscB). This topology was introduced to power laptops as low-voltage and high-current Voltage Regulator Modules as well as non-isolated Point-of-Load converters (Vin < 12 V). To satisfy the converter stability, a state-space modeling technique of switching intervals coupled with parasitic component linearization is developed. Due to the series capacitor charging period miscalculation, the applicability of the ACSM of 2-pscB switching topology for high-power electronic applications is still very limited. Inserting a series capacitor between power switches of phase A increases loop parasitic inductance, introduces a time delay mismatch between the gate voltages of the two switches, and causes interference with the synchronization of the dead time between both phases of 2-pscB converters since the phase B has no series capacitor. This mismatch reduces the heat distribution efficiency and lifetime. As such, a complete model study delivered by the converter is required to design a robust controller. Driven to explore the series capacitor voltage feedback mechanism, frequency analysis of transfer functions, and filter behavior with experimental prototype examples (Vin < 120 V) have been presented for the first time to demonstrate the theoretical analysis. Obtained efficiency was up to 94.9% at full load.

Keywords: two-phase series-capacitor buck converter, state-space averaging, VRMs applications, current sharing mechanism, inductor currents derivation

1. Introduction

The current sharing mechanism concept for conventional multi-phase converters, which are designed based on sensing each phase current to deliver the current information to their controllers, is one of main issues of its implementation. In conventional buck converters, this mechanism may require a preset current sharing ratio at the expense of efficiency, which eventually requires a larger sensing circuit to achieve the sensing accuracy of each phase. Introducing the concept of automatic current sharing in two-phase series capacitor buck topology (2-pscB) is one of the solutions to tackle this issue.

One of the main benefits of a 2-pscB converter is power management to obtain the highest performance of the regulator, automatic power management between phases is used, and the current is evenly distributed between the phases. If the current balance is not fully achieved, input and output ripple cancelation benefits are interrupted, resulting in stability problems. With complete current sharing, better thermal performance and efficiency at higher loads are guaranteed, because all output load is not concentrated in one group of Si MOSFETs/GaNs switches or in a single inductor; otherwise, the unmatched currents may cause a sharp drop in efficiency, instead of two or more phases sharing the thermal burden, the current will pass and cause failure in one phase. The typical topology and time intervals for 2-pscB are shown in **Figure 1**.

There are many advantages attached to this power conversion topology, such as:

- Reducing switching voltage level (e.g., reducing V_{DS} by half). This results in better efficiency when switching losses E_{oss} is reduced and becomes suitable for high-switching frequencies.

- Reducing output voltage ripple and current ripple by the third.

- Reducing the size of the filter inductor by almost half.

- The inductors act as current sources to softly charge and discharge the series capacitor in four-mode power stages (self-charging).

- Flexible intervals time where intervals can (nearly) be reduced to two as a regular buck.

- Automatic current balancing between phases can be easily achieved when the two inductors have the same size, same storage energy, and the inductors' current shapes are the same. In special cases, even if the inductors are not the same, e.g., manipulating the pulse intervals can achieve the current balancing by increasing one or more switch duty cycles.

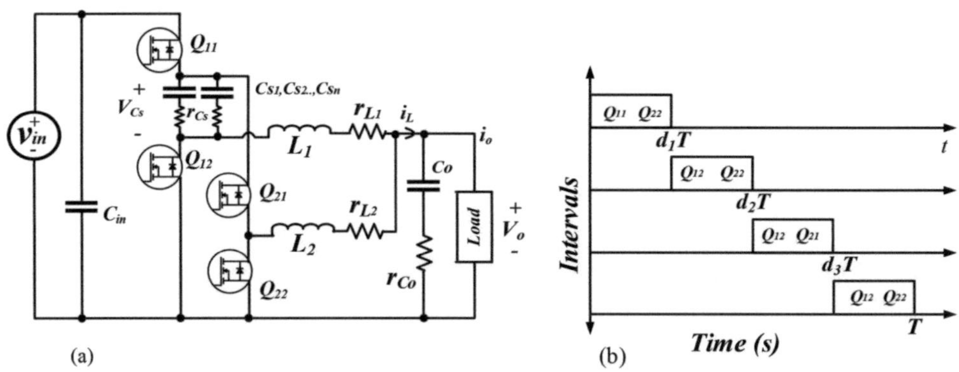

Figure 1.
Two-phase series capacitor buck converter a) topology and b) time intervals.

- In general, there is no need to have any phase current sensing elements to achieve the current sharing, which is a critical issue for distributing the heat, especially at high current demand.
- The duty ratio is doubled to achieve the same specification of a single buck compared with conventional buck converters for the same conversion ratio; this makes series capacitor buck topology easier to control in high-frequency applications.

2. The 2-pscB regulator states equations

The main objective of this section is to derive the transfer functions governing the operation of the 2-pscB converter. Since this converter is nonlinear and time-varying, it utilizes the switching function of the converter power device to achieve high efficiency. Due to the non-ideal characteristics of the switches and their conduction mode resistance during the switching transition, and because the voltage and current will not suddenly become zero during the switching time, this response brings a certain power loss to the system. In modeling, the load current is assumed to be unknown, as such assumed that this topology consists of non-ideal (transistors, inductor, and capacitor), and they have resistance in conduction conditions. The nonlinear—regulator active switches—circuit elements have other non-ideal effects like a voltage drop of conduction mode of active switches, which is neglected due to the complexity of modeling process. This state-space averaging model could be used to design a robust controller that can satisfy the stability and performance conditions of the converter. Small-signal model linearization of four switching intervals using a state-space average model is required. However, the non-idealistic nature of switches with their conduction mode resistance generates some power losses due to the finite voltage and current during switching transitions. Therefore, a complete model includes all the system parameters has to be generated, such as the turn-on resistance of the diode switch, the parasitic resistances of the inductor and capacitor, and the unidentified load current that can be delivered from the converter. The first step in modeling is to convert a complex circuit into a simplified circuit, in which circuit rules can be established. In a switching regulator, a component that stores energy in a circuit or system (such as capacitor voltage and inductor current) is of great significance. The linear and time-invariant system consists of four regions: the two on-regions of input source and series-capacitor energy storage source and two off-regions. The on-time is denoted by D_1T, D_3T, and the off-time is denoted by D_2T, $(1 - D_1\text{-}D_2\text{-}D_3)\ T$.

Thus.

$$
\begin{aligned}
\mathrm{d_1T} &= \mathrm{D_1T}. \\
(d_2 - d_1)\ T &= D_2T. \\
(d_3 - d_2)\ T &= D_3T. \\
(1 - d_3)\ T &= (1 - D_1 - D_2 - D_3)\ T.
\end{aligned}
\tag{1}
$$

In which T is the period of the steady-state output voltage. **Figure 1** shows a two-phase series capacitor buck switching regulator. The four switches are turned on (off) by a pulse with a period of *T*, and its duty cycle is *d* seconds. Therefore, we can represent the simple equivalent circuit of the system in four on and off modes. To look at the stages of operation of this converter in a steady state during the first mode, the high-side switch of phase A, switch Q_{11}, is on, and the inductor current in inductor L_1 charges up the series capacitor a small amount. Using two series capacitors or more Cs_1, Cs_2, .., Cs_n in parallel will reduce the parasitic resistance r_{Cs} to half or less. This topology has a duty cycle as follows.

$$\frac{V_o}{V_{in}} = \frac{d_1(d_3 - d_2)}{d_1 + d_3 - d_2} = \frac{D_1 D_3}{D_1 + D_3} \tag{2}$$

Where d_1, d_2, d_3 represent intervals for capacitor series buck modes.
When these intervals are equal $D_1 = D_2 = D_3 = D = t_{on}/T$.
Hence,

$$\frac{V_o}{V_{in}} = \frac{D}{2} = \frac{t_{on}}{2T} \tag{3}$$

At current sharing balance between the two phases, switching voltage must be the same where,

$$d_1(V_{in} - V_{Cs}) = V_{Cs}(d_3 - d_2) \tag{4}$$

or

$$\frac{V_{Cs}}{V_{in}} = \frac{d_1}{d_1 + d_3 - d_2} = \frac{D_1}{D_1 + D_3} \tag{5}$$

From Eqs. (2, 5) at steady-state condition we conclude that

$$\frac{V_o}{V_{cs}} = D_3 \tag{6}$$

Consideration of i_L, v_{Cs}, and v_{Co} as our state variables of the continuous-time LTI system consists of a state equation and output equation. As a result, the multiphase controller maximizes the duty cycles of both phases during the transient period to the maximum of 25% of the period while maintaining 25% of the time between each duty cycle. All phase inductors that parallel each other are reduced by several phases where a smaller equivalent inductance can charge the output capacitor faster than bigger ones. This issue also reduces the overshooting when the excess charge stored in the inductor of each phase partially discharges at the phases turn-off state, and then the rest transfers to the output capacitor.

On the other hand, the first drawback of this topology is that it has a 50% duty cycle limit. This limitation means that the high-side switches Q_{11} and Q_{21} cannot be turned on at the same time, coupled with the fact that conventional buck gives the converter an inherent 2:1 step-down; therefore, the theoretical minimum input voltage rate is four times the output voltage. In other words, the minimum input voltage is going to be almost five times the output voltage when we take losses into account.

$$V_{in} \geq 4\,(V_o + E_{oss}) \tag{7}$$

3. Small signal average switching model

In literature, many approaches have been proposed for modeling DC-DC converters, e.g., averaged nonlinear formulation in which switching frequency is dependent. Another approach is large-signal presentations of the variable-structure system [1], where the state-space averaging approach is widely used, which yields an average and linearized model formulation [2] depending on the switching frequency, this model is non-linear and time-varying [3]. The sources of disturbance in power DC-DC systems are many [4, 5]. **Figure 2** shows a functional diagram representing some of these sources, where v_o and i_L are dependent on independent inputs.

$$v_o(t) = f_1\{v_{in}(t), i_{out}, D_1(t), D_3(t)\} \tag{8}$$

$$i_L(t) = f_2\{v_{in}(t), i_{out}, D_1(t), D_3(t)\} \tag{9}$$

The equivalent circuit model of the 2-pscB converter can be expressed as seen in **Figure 3**, which contains four independent inputs (input voltage, two control input

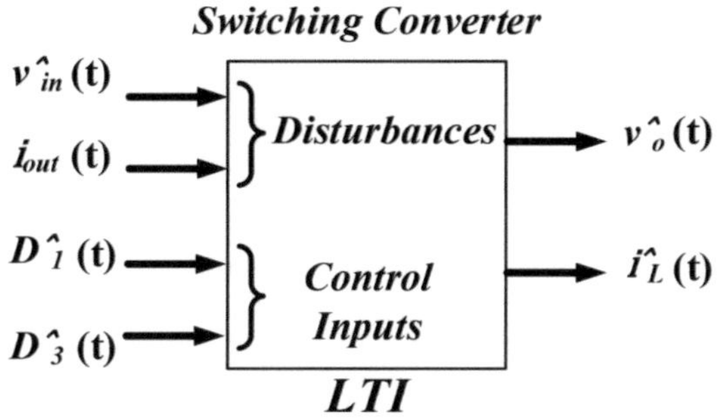

Figure 2.
Block diagram illustrated the dependence of v_o on independent inputs.

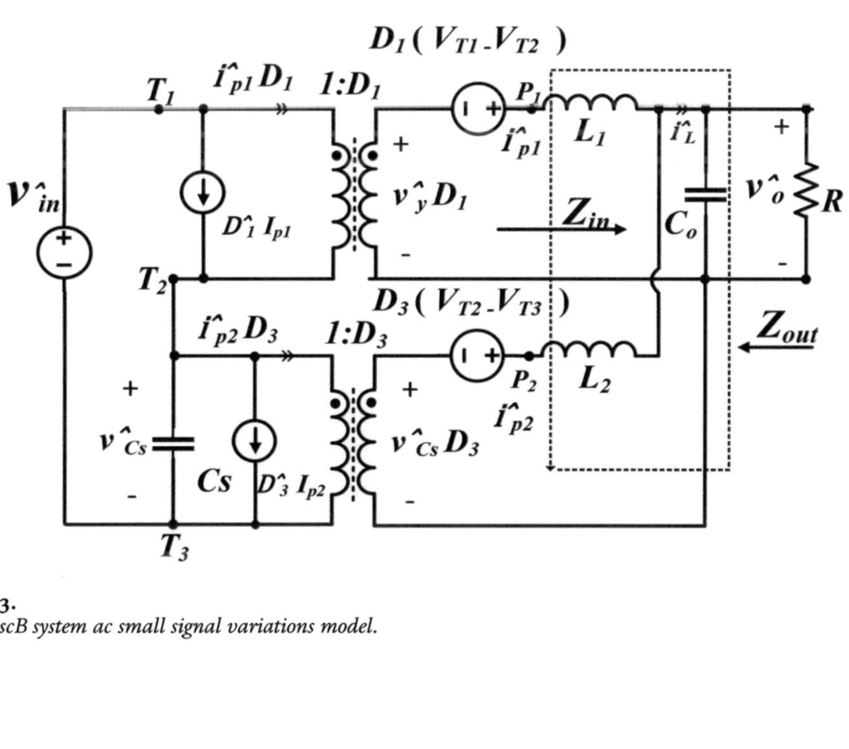

Figure 3.
The 2-pscB system ac small signal variations model.

variation, and load current) and one output dependent variable. Therefore, the ideal transformer averaging model concept can be directly applied to represent the converter.

$$V_{sw1} = D_1(V_{T1} - V_{T2}) = D_1(V_{in} - V_{Cs}) \tag{10}$$

$$V_{sw2} = D_3(V_{T2} - V_{T3}) = D_3 V_{Cs} \tag{11}$$

By realizing Eqs. (10) and (11), a large signal average switch model can be formed with transformers and similar average current equations, hence,

$$v_y = v_{in} - v_{Cs} \tag{12}$$

$$i_L(t) = i_{P1}(t) + i_{P2}(t) \tag{13}$$

$$\frac{v_o}{i_L} = R//\frac{1}{SC_o} \tag{14}$$

where p_1, p_2 are phase current points, at complete current sharing conditions, ideal transformers for both phases will be symmetrical such as

$$D_1 \hat{i}\, p_1 = D_3 \hat{i}\, p_2 \tag{15}$$

Another way to illustrate 2-pscB converter system ac small signal variations model is seen in **Figure 4**, the output voltage variation can be expressed.

$$v_o(s) = G_{vd}(s)\{D_1(s) + D_3(s)\} + G_{v-in}(s)v_{in}(s) - Z_{out}(s)i_{out}(s) \tag{16}$$

Where $G_{v-in}(s)$ and $G_{vd}(s)$ are the line-to-output and control-to-output transfer function expression, respectively. Manipulate block diagram to solve for $v_o(s)$. Hence

$$v_o = V_{ref}\frac{G_c G_r G_{vd}/V_M}{1 + HG_c G_r G_{vd}/V_M} + V_{in}\frac{G_{v-in}}{1 + HG_c G_r G_{vd}/V_M} \pm i_{out}\frac{Z_{out}}{1 + HG_c G_r G_{vd}/V_M} \tag{17}$$

which is of the form,

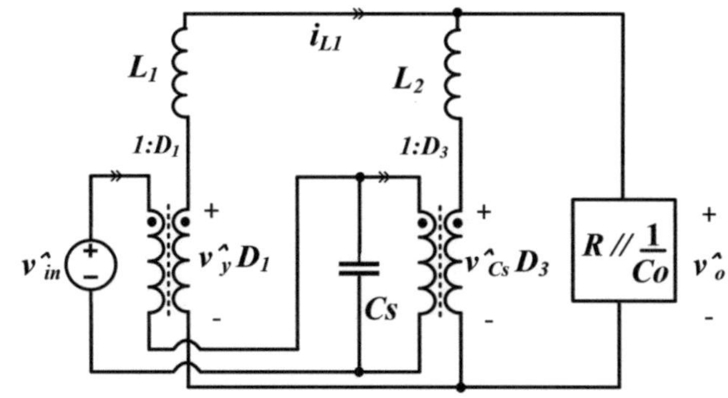

Figure 4.
Open loop impedance of 2-pscB converter.

$$v_o = V_{ref} \frac{1}{H} \frac{T}{1+T} + V_{in} \frac{G_{v-in}}{1+T} \pm i_{out} \frac{Z_{out}}{1+T} \tag{18}$$

With $T(s) = H(s)G_c(s)G_r(s)G_{vd}(s)/V_M$

$T(s)$ is the product of the small-signal gains in the foreword and feedback paths of the control loop, the modulator voltage is V_M, and $H(s)$ is the current sensor gain. Z_{out} is converter output impedance formula. To find the output voltage at the equilibrium case and complete current sharing, writing the *KVL* for the simple loops of **Figure 5**, we get:

$$v_o = i_L \left(R // \frac{1}{sC_o} \right) = (i_{L1} + i_{L2}) \left(R // \frac{1}{sC_o} \right) \tag{19}$$

$$i_{L1}(2sL) - i_{L2}(sL) = \frac{1}{2} v_{in} D_1 + \frac{1}{2} v_{in} D_3 \tag{20}$$

$$i_{L2}(sL) - i_{L1}(sL) + i_{L2} \left(R // \frac{1}{sC_o} \right) = \frac{1}{2} v_{in} D_3 \tag{21}$$

After calculating the previous equations, we conclude

$$v_o = \frac{2R}{sL(sRC_o + 1)} \left(\frac{1}{4} v_{in}(D_1 + D_3) - v_o \right) \tag{22}$$

In other words,

$$\frac{v_o}{v_{in}} = \frac{\frac{D_1 + D_3}{4}}{s^2 \frac{L\,C_o}{2} + s\,\frac{L}{2R} + 1} \tag{23}$$

This formula represents converter line-to-output transfer function.

$$G_{v-in}(s) = \left. \frac{v_o(s)}{v_{in}(s)} \right|_{D_1 = D_3 = 0, i_{out} = 0} = \frac{\frac{D_1 + D_3}{4}}{\frac{s^2}{\omega_o^2} + \frac{s}{Q_o \omega_o} + 1} \tag{24}$$

At $D_1 = D_3 = D$

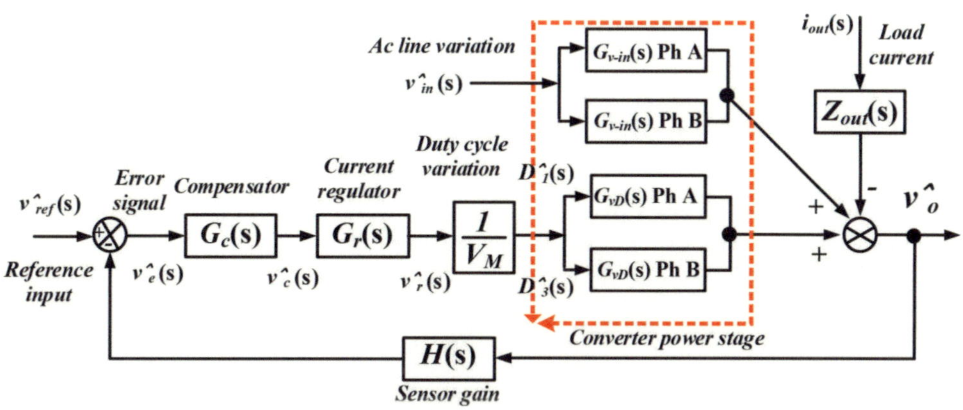

Figure 5.
Complete block diagram of two-phase series capacitor voltage regulator.

$$G_{v-in}(s) = \frac{D}{2} = G_{go}$$

And

$$G_{v-in}(s) = G_{v-in}(s)|_{phase\ I} + G_{v-in}(s)|_{phase\ II}$$

$$= \frac{v_o(s)}{v_{in}(s)}|_{D_1=0,i_{out}=0} + \frac{v_o(s)}{v_{in}(s)}|_{D_3=0,i_{out}=0} \tag{25}$$

$$G_{v-in}(s)|_{phase\ I} = G_{v-in}(s)|_{phase\ II} = \frac{\frac{D_{1,3}}{4}}{\frac{s^2}{\omega_o^2} + \frac{s}{Q_o\omega_o} + 1} \tag{26}$$

Where the angular frequency and quality factor are

$$\omega_o = \sqrt{\frac{2}{C_oL}}, Q_O = R\sqrt{\frac{2\,C_o}{L}}$$

To generate a Laplacian formula represents converter control-to-output transfer function $G_{vD}(s)$,

$$G_{vD}(s) = \left.\frac{v_o(s)}{D(s)}\right|_{v_{in}=0,i_{out}=0} = \frac{\frac{v_{in}}{2}}{\frac{s^2}{\omega_o^2} + \frac{s}{Q_o\omega_o} + 1} \tag{27}$$

$$G_{vD}(s) = G_{vD}(s)|_{phase\ I} + G_{vD}(s)|_{phase\ II} = \left.\frac{v_o(s)}{D_1(s)}\right|_{v_{in}=0,i_{out}=0} + \left.\frac{v_o(s)}{D_3(s)}\right|_{v_{in}=0,i_{out}=0} \tag{28}$$

At complete current sharing

$$G_{vD}(s)|_{phase\ I} = G_{vD}(s)|_{phase\ II} = \frac{\frac{v_{in}}{4}}{\frac{s^2}{\omega_o^2} + \frac{s}{Q_o\omega_o} + 1} \tag{29}$$

To derive converter output impedance formula

$$Z_{out} = \left.\frac{v_o(s)}{i_{out}(s)}\right|_{v_{in}=0,D_1=D_3=0}$$

$$= \left(R//\frac{sL}{2}//\frac{1}{sC_o}\right) = \frac{2\,sL}{s^2\frac{L\,C_o}{2} + s\,\frac{L}{2R} + 1} = \frac{2\,sL}{\frac{s^2}{\omega_o^2} + \frac{s}{Q_o\omega_o} + 1} \tag{30}$$

In some analyses, accurate models are needed; thus, these formulas represent the transfer functions with output filter parasitic components,

$$G_{v-in}(s) = \left.\frac{v_o(s)}{v_{in}(s)}\right|_{D_1=D_3=0,i_{out}=0}$$

$$= \frac{\frac{D}{2}\left(s\frac{2C_or_{Co}}{2R+r_L} + \frac{2R}{2R+r_L}\right)}{s^2\left(\frac{RLC_o+LC_or_{Co}}{2R+r_L}\right) + s\left(\frac{L+RC_or_L+C_or_Lr_{Co}+2RC_or_{Co}}{2R+r_L}\right) + 1} \tag{31}$$

$$G_{vD}(s) = \left.\frac{v_o(s)}{D(s)}\right|_{v_{in}=0, i_{out}=0}$$

$$= \frac{v_{in}\frac{R}{2R+r_L}\left(s\frac{C_o r_{Co}}{R}+1\right)}{s^2\left(\frac{RLC_o+LC_o r_{Co}}{2R+r_L}\right)+s\left(\frac{L+RC_o r_L+C_o r_L r_{Co}+2RC_o r_{Co}}{2R+r_L}\right)+1}$$

$$= \frac{R}{2R+r_L}\frac{v_{in}\left(\frac{s}{\omega_x}+1\right)}{\frac{s^2}{\omega_o^2}+\frac{s}{Q_o\omega_o}+1} \tag{32}$$

Where,

$$\omega_x = \frac{R}{C_o r_{Co}}, \omega_o = \sqrt{\frac{2R+r_L}{LC_o(R+r_{Co})}}, Q_o = \frac{\sqrt{LC_o(2R+r_L)(R+r_{Co})}}{(L+RC_o r_L+C_o r_{Co}\, r_L+2RC_o r_{Co})} \tag{33}$$

Therefore, the choice of components used in a switching regulator has a large impact on its performance. Critical components such as switching elements, magnetic components, and filter capacitors all affect both the switching frequency and the overall efficiency of the converter. In the previous sections, the power switch, inductances, and capacitances were all considered ideal components. But real components are not ideal and have parasitic properties, which will affect the overall performance of the DC-DC converter. A typical output capacitor always exhibits stray elements such as r_{Co}. This parasitic element introduces a zero in the control-to-output transfer function see Eq. (32). The relationship between the output parasitic capacitor (r_{Co} + Δr_{Co}) and converter control-to-output transfer function is illustrated in **Figure 6**.

When using the ceramic output capacitor, the output capacitor parasitic r_{Co} effect is aggravated at relatively high frequencies, often above the switching frequency so it can be seen in the gain graph. The parameter ω_o is the angular corner frequency, which is defined as follows.

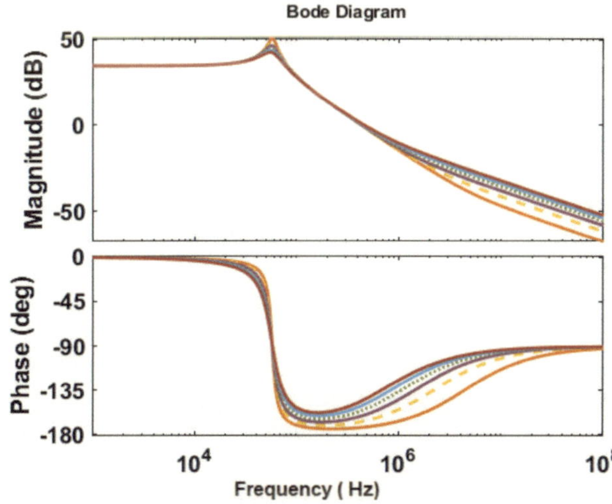

Figure 6.
G_{vD} and output capacitor ESR parasitic variation from 0 to 90 mΩ.

$$f_o = \frac{\omega_o}{2\pi}$$

$$f_o = \frac{1}{2\pi}\sqrt{\frac{2R + r_L}{RLC_o + LC_o r_{Co}}} \tag{34}$$

To get a well-regulated average output voltage signal, the switching frequency must be greater than 10 times of angular corner frequency (cutoff frequency) for small or invisible output voltage ripple.

$$f_{sw} > 10 f_o$$

Thus, $f_{sw} \approx 20 f_o$ is recommended. The converter output impedance transfer function formula, including output filter parasitic, is presented as follows.

$$Z_{out} = \frac{s^2\left(\frac{RLC_o r_{Co}}{2R + r_L}\right) + sR\left(\frac{L + C_o r_{Co} r_L}{2R + r_L}\right) + \frac{R\, r_L}{2R + r_L}}{s^2 LC_o\left(\frac{R + r_{Co}}{2R + r_L}\right) + s\left(\frac{L + RC_o r_L + C_o r_L r_{Co} + 2RC_o r_{Co}}{2R + r_L}\right) + 1} \tag{35}$$

For inequality conditions, we find

$$Z_{out} = \frac{Z_g\left(\frac{S^2}{\omega_k^2} + \frac{S}{Q_k \omega_k} + 1\right)}{\left(\frac{S^2}{\omega_o^2} + \frac{S}{Q_o \omega_o} + 1\right)} \tag{36}$$

Where:

$$Z_g = \frac{Rr_L}{2R + r_L},\ \omega_k = \sqrt{\frac{r_{L1} + r_{L2}}{(L_1 + L_2)C_o r_{Co}}},\ Q_k = \frac{\sqrt{(r_{L1} + r_{L2})(L_1 + L_2)C_o r_{Co}}}{(L_1 + L_2) + (r_{L1} + r_{L2})C_o r_{Co}} \tag{37}$$

$$\omega_o = \sqrt{\frac{2R + r_L}{LC_o(R + r_{Co})}},\ Q_o = \frac{\sqrt{LC_o(2R + r_L)(R + r_{Co})}}{(L + RC_o r_L + C_o r_{Co}\, r_L + 2RC_o r_{Co})} \tag{38}$$

From Eq. (35), inductance parasitic variation ($r_L + \Delta r_L$) affects output impedance in lower frequencies, often below the switching frequency so the gain graph increases with the inductance parasitic increases, as can be seen in **Figure 7**.

Output capacitor ESR parasitic variation affects the output impedance in high frequencies, as can be seen in **Figure 8**. From the previous two figures, we can summarize the observations in another way, where the output impedance is represented by the contribution of the filter components, as shown in **Figure 9**. By the below graph's inspection of **Figure 9**, we can see that the inductor resistive path r_L dominates the impedance in DC. As frequency increases, the inductor then enters the spectrum. The capacitor impedance starts to take over the inductive section at higher frequencies until it becomes a short circuit and leaves the output impedance value to its series loss r_{Co}. To solve for the peak value of output impedance at resonance value of f_o, since r_{Co} contribution is small and neglected at low frequencies.

$$|Z_{out-Max}|dB = R\sqrt{\frac{2\,Z_o^2 + r_L^2}{\frac{2}{Z_o^2}\left(Z_o^2 + Rr_L\right)^2 + r_L^2}} \tag{39}$$

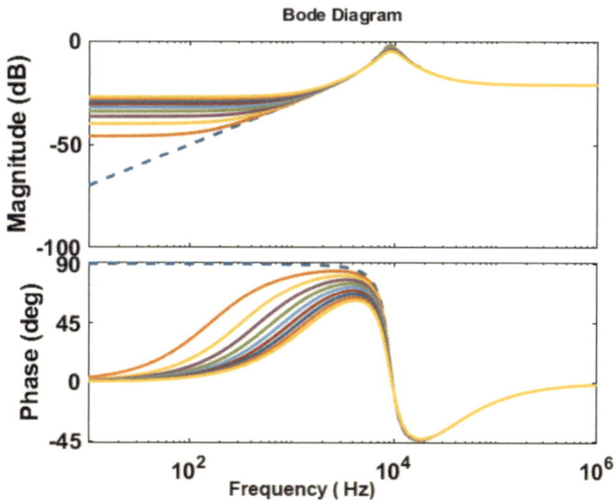

Figure 7.
Z_{out} and inductor parasitic r_L variation from 0 to 90 mΩ..

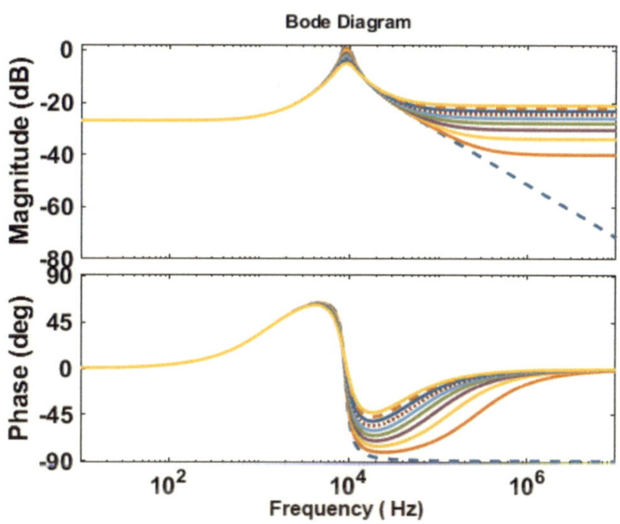

Figure 8.
Z_{out} and output capacitor ESR parasitic variation from 0 to 90 mΩ..

Where $Z_o = \sqrt{\frac{L}{C_o}}$ is the characteristic impedance of the filter.

The output filter size needs to be very carefully selected to minimize voltage drop and power loss, which can be achieved by minimizing the output impedance [6]. Therefore, to get rid of the resonance frequency and maintain a good gain value, it is necessary to select a natural frequency higher than the resonance frequency. The natural frequency should be in the output capacitor region, where the influence of inductance is minimal. Thus,

$$|Z_{o-Min}| \cong \sqrt{\left(\frac{1}{2\,\pi f_{max}\,C_o}\right)^2 + r_{Co}^2} \tag{40}$$

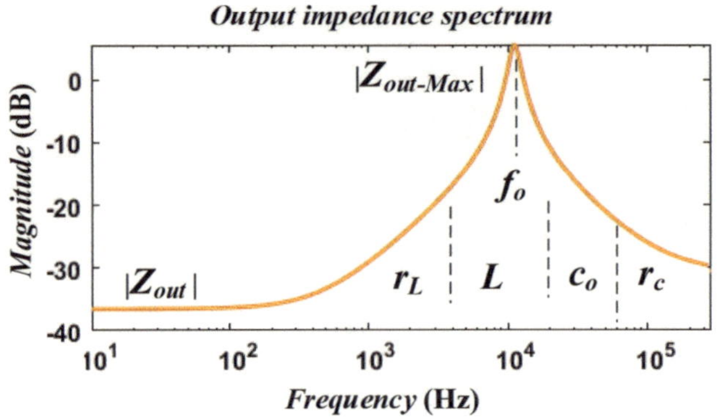

Figure 9.
Output impedance bode diagram spectrum.

4. Inductor currents derivation

Based on the concept of automatic current sharing, we need to study the inductor current of each phase. For steady-state conditions, as seen from **Figure 10**, during interval state one and state three, the current through the inductors increases in phase sequence because energy is being stored in the inductors from the input supply or series capacitors. During the off-time of each phase, the current through inductors decreases as both inductors are sourcing energy to the output. Note that the current increase at state one is equal to the decrease in current during state two and state three of the same phase. The average series capacitor voltage v_{Cs} maintains its value at exactly half of the input voltage with a small voltage ripple, thus it can be regarded as a constant voltage source. A large magnitude difference in the inductor currents with 180° out of phase can be seen during the initial transient. This magnitude difference decreases as the steady-state condition is attained. The duration of the transient time

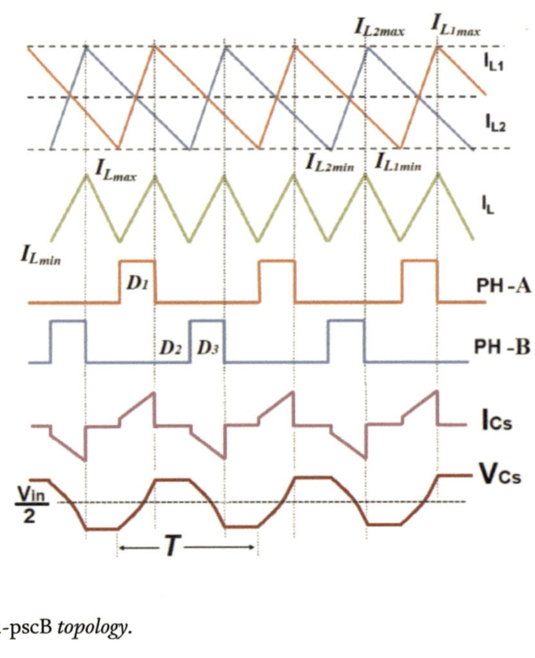

Figure 10.
The main waveforms of 2-pscB *topology.*

depends on dead time matching between the two phases and can be reduced using a soft start scheme with a pre-charged series capacitor. However, this will introduce a delay time to the circuit. The output voltage ripple can be controlled by the size of the output filter. The average charging and discharging currents for the series capacitor are similar, this verifies the steady-state operation of the 2-pscB converter. From this figure, the current ripple (Δi_{L_1}) through the phase inductor is greater than the total output current ripple (Δi_L).

State four is a repeat of state two with both inductors connected to the ground and supplying energy to the output capacitor. In state one, when the average inductor current of phase A is larger than the average inductor current of phase B, the series capacitor voltage v_{Cs} would slowly increase. For VRMs applications, most processors recommend low-output voltage ripple where these applications demand keeping the output current ripple low. This implies the need for a large inductor, the relationship expressed as follows:

$$V_{ripple} = I_{ripple} * r_{Co}, I_{ripple} \propto \frac{1}{L} \tag{41}$$

On the contrary, as mentioned before, the other requirement is the fast-transient response. This justifies the need for a small inductor to allow the current through the supply to change quickly, but at the same time, this assumption conflicts directly with the need for a larger inductor to minimize output voltage ripple.

A mathematical representation of the 2-pscB converter with its internal current sharing mechanism resulting from the v_{Cs}. **Figure 11** shows the average model of the 2-pscB converter, including the main parasitic components and the CSM using the internal series capacitor voltage feedback loop. This block diagram aims to generate a reference trajectory to be used by the controller model.

Where the two phases' currents are:

$$i_{L1}(t) = \frac{1}{L_1}\int_0^{\infty} (V_{in}D_1 - V_{Cs}D_1 - V_o(t))dt \tag{42}$$

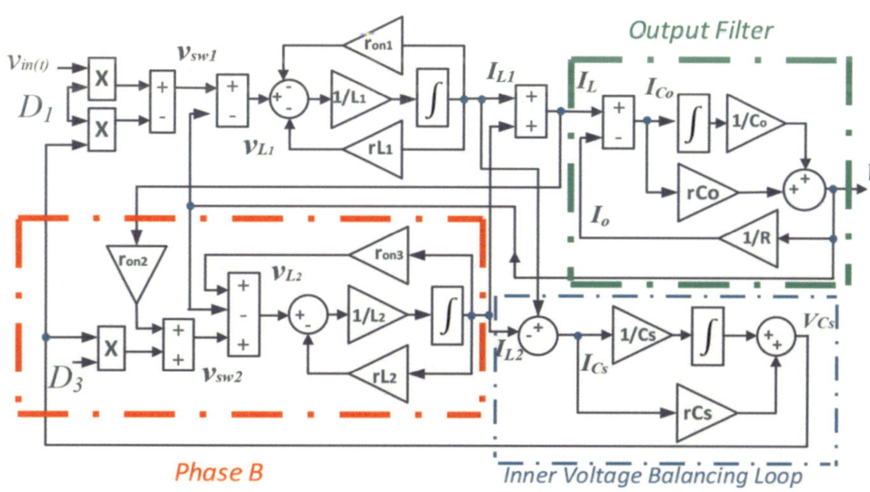

Figure 11.
The 2-pscB *converter average model diagram of inductor currents and current sharing mechanism.*

$$i_{L2}(t) = \frac{1}{L_2}\int_0^{\infty} (V_{Cs}D_3 - V_o(t))dt \tag{43}$$

From current (Eqs. 42, 43), the first derivative of two-phase currents is:
At $D_1 = D_3 = D$ and $L_1 = L_2 = L$

$$\frac{di_L}{dt} = \frac{di_{L1}}{dt} + \frac{di_{L2}}{dt} \tag{44}$$

$$= \frac{1}{L}V_{in}D - \frac{2}{L\,Co}\int_0^{T}\left(i_L - \frac{V_o(t)}{R}\right)dt \tag{45}$$

The second derivatives of two-phase inductor currents, where

$$\frac{di_{L1}^2}{dt^2} = -i_{L1}\frac{1}{L}\left(\frac{D}{Cs} + \frac{1}{Co}\right) + i_{L2}\frac{1}{L}\left(\frac{D}{Cs} - \frac{1}{Co}\right) + \frac{V_o}{L\,Co\,R} = a_1\,i_{L1} + b_1 i_{L2} + c_1 \tag{46}$$

$$\frac{di_{L2}^2}{dt^2} = i_{L1}\frac{1}{L}\left(\frac{D}{Cs} + \frac{1}{Co}\right) - i_{L2}\frac{1}{L}\left(\frac{D}{Cs} - \frac{1}{Co}\right) + \frac{V_o}{L\,Co\,R} = a_2\,i_{L1} + b_2 i_{L2} + c_2 \tag{47}$$

Using Laplace transform, we can derive the current time domain formulas as

$$\begin{bmatrix} i''_{L1} \\ i''_{L2} \end{bmatrix} = \begin{bmatrix} a_1 & b_1 \\ a_2 & b_2 \end{bmatrix}\begin{bmatrix} i_{L1} \\ i_{L2} \end{bmatrix} + \begin{bmatrix} c_1 \\ c_2 \end{bmatrix} \tag{48}$$

For initial conditions of $i_{L1}(0) = 0$, $i_{L2}(0) = 0$, and.
$\frac{di_{L1}(0)}{dt} = \frac{v_{in}}{L}D, \frac{di_{L2}(0)}{dt} = 0$

$$\frac{di_L^2}{dt^2} = \frac{2}{L\,Co}\left(\frac{V_o(t)}{R} - i_L\right) = -\frac{2}{L\,Co}i_{Co} \tag{49}$$

Since $V_{in} = 0$ for the second derivatives, the average output capacitor current equals zero.

$< i_{Co} > = 0$, Then

$$\frac{di_L^2}{dt^2} = 0.$$

5. Series capacitor damping behavior

To examine the damping oscillation of the current sharing of the two phases regarding inductor resistive parasitic variation. Back to Eq. (42), we can express its formula as:

$$L_1\frac{di_{L1}(t)}{dt} = V_{in}D_1 - i_L r_{on1} - V_{Cs}D_1 - i_{L1}r_{Cs} - i_{L1}r_{L1} - V_o(t) \tag{50}$$

$$L_2\frac{di_{L2}(t)}{dt} = -i_L r_{on2} + V_{Cs}D_3 - i_{L2}r_{Cs} - i_{L2}r_{on3} - i_{L2}r_{L2} - V_o(t) \tag{51}$$

Where r_{on1}, r_{on2}, r_{on3}, and r_{on4} represent converter switches on-state resistances, and and r_{Cs} represents the total series capacitor resistive parasitic component.

Where

$$V_{Cs}(t) = \frac{1}{Cs}\int_0^{\infty} (i_{L1} - i_{L2})\,dt \tag{52}$$

Plugging Eq. (52) and substituting Eq. (50) into Eq. (51), the results at full current sharing ($r_{on} = r_{on1} = r_{on2} = r_{on3} = r_{on4}$, and $r_L = r_{L1} = r_{L2}$) will be,

$$\frac{d^2 i_{Cs}(t)}{dt^2} + \frac{1}{L}\left(\frac{di_{Cs}(t)}{dt}(r_L + r_{on} + r_{Cs}) + \frac{(D_1 + D_3)}{Cs} i_{Cs}\right) = 0 \tag{53}$$

To analyze the effect of r_L, assume that other parasitic parameters are very small or zero (ideal conditions) to simplify the results also for homogeneous second-order differential equation,

$$\frac{d^2 i_{Cs}(t)}{dt^2} + \frac{1}{L}\left(\frac{di_{Cs}(t)}{dt} r_L + \frac{(D_1 + D_3)}{Cs} i_{Cs}\right) = 0 \tag{54}$$

Eq. (54) can be rearranged to

$$\frac{d^2 i_{Cs}(t)}{dt^2} + 2\zeta\omega_{Cs}\frac{di_{Cs}(t)}{dt} + \omega_{Cs}^2 i_{Cs} = 0 \tag{55}$$

which is in the form of a second-order differential equation, representing the damped harmonic oscillator nature with attenuation ($\zeta\omega_{Cs}$) and the angular resonant frequency of (ω_{Cs})

$$\zeta = \frac{r_L}{2}\sqrt{\frac{Cs}{L\,(D_1 + D_3)}},,, \omega_{Cs} = \sqrt{\frac{(D_1 + D_3)}{L\,Cs}},\ Q_{Cs} = \frac{1}{r_L}\sqrt{\frac{L\,(D_1 + D_3)}{Cs}} \tag{56}$$

Setting initial values $i_{Cs}\ (0^+) = 0$, From Eq. (50)

$$\frac{di_{Cs}(0)}{dt} = c_s\frac{d^2 v_{Cs}(0)}{dt^2} = \frac{D}{L}(\,V_{in} - 2v_{Cs}(0)) \tag{57}$$

thus,

$$v_{Cs}(0^+) = V_{Cs} - \frac{\Delta V_{Cs}}{2} \tag{58}$$

The damped harmonic oscillator can be translated as current perturbation representing the difference in phases average inductor currents. The effect of series capacitor ESR and switches internal on-state resistance is similar to the inductor DCR effect. For different parameters settings, to find the solution of second-order, linear, homogeneous differential equations with constant coefficients, there is a need to figure out the characteristic equation as.

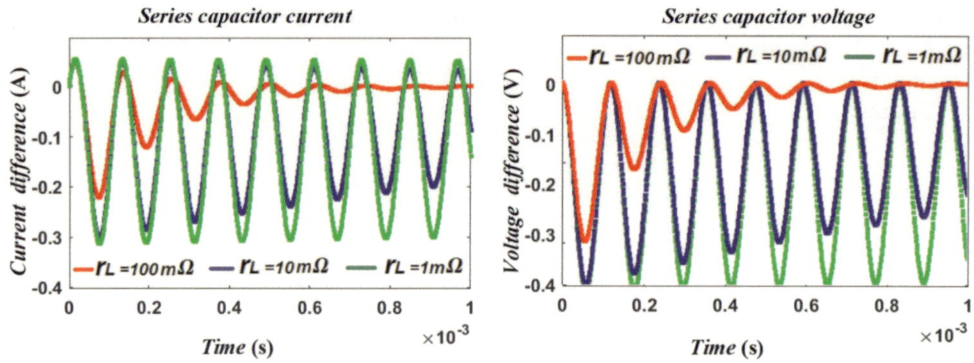

Figure 12.
Series capacitor characteristic equation response of (a) the difference in average inductor currents due to current perturbation in inductor currents for varying values of lumped inductor resistance and (b) the difference in average series capacitor voltage.

$$a^2 + P_1 a + P_0 = 0 \tag{59}$$

To solve equations With real coefficients, the complex roots are always distinct if they are not purely real. So, for complex roots $s_{1,2} = \alpha \mp j\beta$

$$i_{Cs} = \lambda_1 e^{\alpha t} cos\beta t + \lambda_2 e^{\beta t} sin\beta t \tag{60}$$

Where λ_1 and λ_2 are arbitrary constants, **Figure 12** depicts this type of solution.

6. Unbalanced series capacitor voltage

The 2-pscB converter current sharing mechanism tolerance will be discussed here, for many reasons, as the case of unmatched duty cycles, this topology has the ability to maintain adequate current sharing balance for a certain limit. This limit depends on the charging time of the series capacitor, which in general is affected by the quality of series capacitor material.

Since

$$\frac{V_{Cs}}{V_{in}} = \frac{D_1}{D_1 + D_3}$$
$$V_{sw1} = V_{in} - V_{Cs}$$
$$V_{sw2} = V_{Cs}$$
$$\frac{V_{sw1}}{V_{sw2}} = \frac{D_3}{D_1} = \frac{d_3 - d_2}{d_1} \tag{61}$$

It turns out that under phase unbalanced conditions (duty cycle mismatch, phase placement issues, etc.), the transient time is longer than the transient time under balanced conditions when $D_1 = D_3$. For unbalanced conditions, the areas of the switching voltage node pulses for both phases remain the same for a certain tolerance or boundary.

$$V_{sw1}{}^{*} D_1 = V_{sw2}{}^{*} D_3 = V_o \tag{62}$$

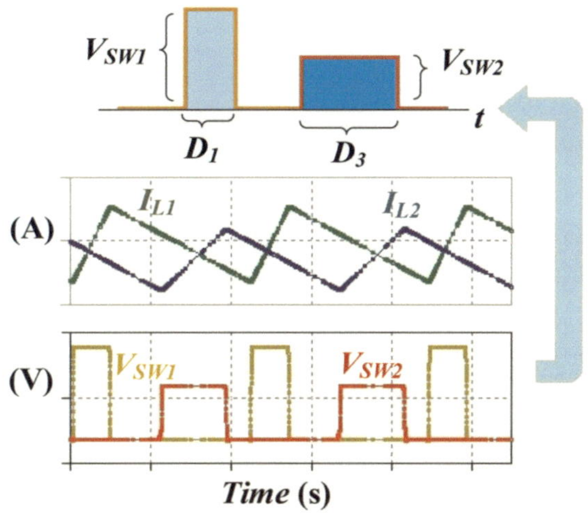

Figure 13.
The 2-pscB converter current sharing mechanism tolerance at $D_1 = 0.2$, $D_3 = 0.35$.

An example to present these conditions is, at $D_1 = 0.2$ and $D_3 = 0.35$ or opposite values, as shown in **Figure 13**,

$$\frac{V_{sw1}}{V_{sw2}} = \pm 1.75 \text{ and } V_{sw1} \neq V_{sw2} \tag{63}$$

Therefore, phase A switching node voltageV_{sw1} increased to compensate for the reduction in the on-time of the phase. As shown in **Figure 13**, in both buck phases, when the high-side switch is turned off, the low-side (synchronous rectifier) switch is turned on, and the current is circulating through the lower switch. Since the inductor current cannot instantaneously stop, the "on-time" and "off-time" of the switch are unbalanced. Each phase inductance is large enough to ensure that it works in the continuous conduction mode (CCM). For certain limits, during steady-state operation, the series capacitor voltage can still manage to maintain the desired output voltage and stabilize current sharing among the two phases [7–12]. The 2-pscB topology has been fully investigated using different Si MOSFETs and GaN switches. The theoretical and experimental testing specifications for the eGaN-based 48 V/5Vdc 2-pscB converter are listed in **Table 1**.

Figure 14 shows the main proposed prototypes and evaluation modules of GaN transistors and Si MOSFETs 2-pscB converters at different voltage rate (V_{in} < 120 V). Experimental results to examine the CSM for the phase-sensorless GaN-based 48 V/ 5 V 2-pscB converter with an efficiency of up to 94.9% are shown in **Figure 15**. To construct the converter GS61004B transistors with $R_{ds(on)}$ of 15 mΩ are used. Despite the two-phase parasitic components' layout unbalance, the waveform shows output voltage, series capacitor current, and the fully automatic current balancing (≈100%) between two phases.

The output current ripple (I_L) is less than the phase ripple ($i_{L(rms)} = i_{L1(rms)} + i_{L2(rms)}$), which is one of the technical advantages of multi-phase that the combined output ripple (total ripple) is less than the ripple current in each phase. This occurs due to driving the phases out of phase. To achieve the complete current sharing

Description	Label	Parameter
Series capacitor	Cs	< 4.7μF
Phase inductors	L_1,L_2	4.7μH
Output capacitor	Co	>70μF
Switching frequency	f_{sw}	250 kHz
Input voltage	V_{in}	48 V
Output voltage	V_o	5-5.5 V
Cs stray component	r_{Cs}	10 mΩ
Switches stray components	r_{on},r_{L1},r_{L2}	< 20 mΩ

Table 1.
2-pscB converter evaluation parameters.

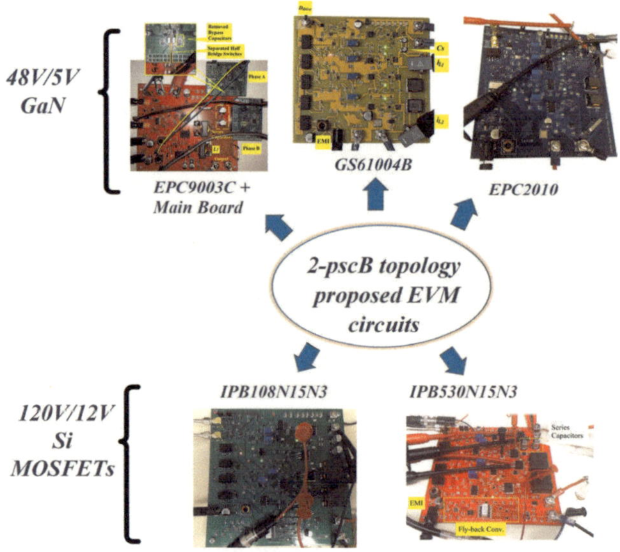

Figure 14.
The proposed 2-pscB *converter EVM prototypes using different switches at* (V_{in} < 120 V).

balance between two phases, the switching nodes voltage must be the same. Also, for the voltage equations of phase inductors v_{L1} and v_{L2} (see **Figure 1a**), these voltages must equal each other. This equivalency gives the expression of the series capacitor voltage in terms of the topology parasitic values as follows:

$$v_{Cs} = v_{in}\frac{D_1}{D_1 + D_3} + \frac{i_{L_1}\in_1 + i_{L_2}\in_2}{D_1 + D_3} \tag{64}$$

Where parasitic coefficients are

$$\in_1 = r_{on}(1 + D_3 - 2D_1) - D_1 r_{Cs} - r_{L1}$$
$$\in_2 = 2D_3 r_{on} + D_3 r_{Cs} - r_{on} + r_{L2}$$

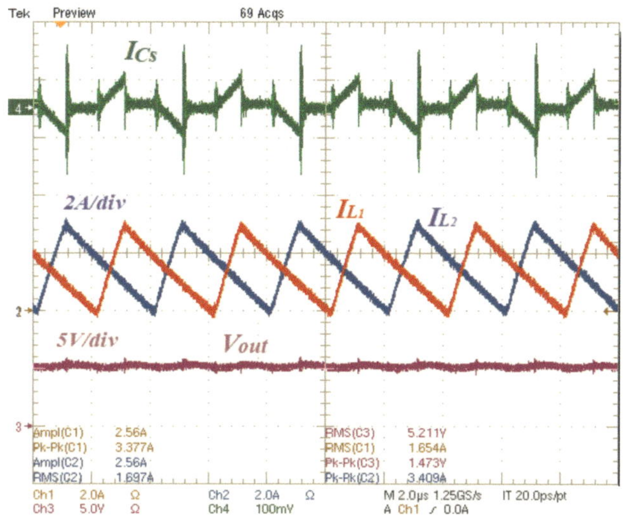

Figure 15.
The 2-pscB *converter typical waveforms of* i_{Cs} (above), i_{L1}, i_{L2} (middle), and V_o *(below) at 48 V input and 250 kHz.*

For small variations in topology duty cycles, parasitic coefficients can be neglected. The complete current sharing conditions can be reached at $\in_1 = \in_2 \cong (0)$

Analytical and experimental comparisons between the traditional buck topology and the 2-pscB topology show that the 2-pscB topology can reduce power losses by up to 31% at full load. Analytical estimates show that the maximum converter inductor size reduces by 55%, and inductor current ripple reduces by 30% compared with conventional bucks at the same switching frequency and duty cycle. This study concludes that the 2-pscB topology requires only 34% of the conventional buck capacitor [13–16]. Please note that the efficiency of this topology increases with the size of the inductor, but at the cost of transient response performance.

7. Conclusion

The main purpose of this chapter is to provide a comprehensive overview of the 2-pscB topology current sharing mechanism starting from the model of the AC small-signal variation of the system to the characteristic equations response of balanced and unbalanced series capacitor voltage. This would help in the enhancement of the overall performance. Theoretical analysis of damping behavior of series capacitor current is formulated, and series capacitor voltage compensation mechanism of the topology current was presented. Detailed studies of the mathematical representation modeling and Laplacian matrices of the 2-pscB converter with an internal voltage feedback loop were presented to help build an efficient converter controller. Based on the above, this topology needs an additional control circuit that can limit the inrush current caused by series capacitance and output capacitor charging period at the beginning of the transient time. The duration of the transient time depends on dead time matching between the two phases and can be reduced using a soft start scheme with a pre-charged series capacitor. Since phase B has no series capacitor, for traditional half-bridge gate driving circuits, inserting a series capacitor between power switches of phase A increases loop parasitic inductance, introducing a time delay mismatch between the gate voltages of the phase switches. This mismatch can eventually cause interference with the synchronization of the dead time between phase A and B of 2-pscB converters. Using traditional half-bridge gate driving circuits in 2-pscB, the turn-off delay of phase A switches caused by its C_{DS} discharging period appears and amplifies by parasitic effects and the phases' sinking path mismatch. Careful consideration has to be given to the specifications of the series capacitor to tackle parasitic and mismatch problems.

Conflicts of interest

The author declares that there are no conflicts of interest regarding the publication of this chapter.

Author details

Salahaldein A. Rmila
Department of Electrical Engineering, University of Arkansas, Fayetteville, USA

*Address all correspondence to: saa016@uark.edu

References

[1] Mahdavi J, Emaadi A, Bellar M. Analysis of power electronic converters using the generalized state-space averaging approach. IEEE Transactions on Circuits and Systems I. 1997;**44**(8): 767-770

[2] Perreault DJ, Verghese GC. Time-varying effects in models for current-mode control. In: Proceedings of PESC '95—Power Electronics Specialist Conference. Vol. 1. 1995. pp. 621-628

[3] Ridley RB. A new, continuous-time model for current-mode control. IEEE Transactions on Power Electronics. 1991; **6**(2):271-280

[4] Erickson RW, Maksimovi D. Fundamentals of Power Electronics. USA, NY: Springer-Verlag; 2001. pp. 589-607

[5] Zumbahlen H. Basic linear design, analog devices. In: Linear Circuit Design Handbook. Boston: Elsevier-Newnes; 2006

[6] Basso C. Switch-Mode Power Supplies: SPICE Simulations and Practical Designs. 2nd ed. NY: McGraw Hill; 2014

[7] Rmila SA, Chen Z. A comparison between conventional Buck and 2-pscB DC-DC converters. IEEE Texas Power and Energy Conference (TPEC). 2021; **2021**:1-6

[8] Ahmed S, Ang SS. A high-input voltage two-phase series capacitor DC-DC Buck converter. Journal of ECE. 2020;**2020**:Article ID 9464727

[9] Rmila SA, Chen Z. Investigating the CSM of a GaN-based two-phase series capacitor Buck DC-DC converter. IEEE Power and Energy Conference at Illinois (PECI). 2022;**2022**:1-7

[10] Shenoy PS, Amaro M, Freeman D, Morroni J. Comparison of a 12V, 10A, 3MHz buck converter and a series capacitor buck converter. In: IEEE Applied Power Electronics Conf. and Exposition (APEC). Vol. 2015. Charlotte, NC: APEC; 2015. pp. 461-468

[11] Ahmed S. Analysis and Optimization of the Two-Phase Series Capacitor Buck DC-DC Converter. Arkansas: University of Arkansas; 2022

[12] Wang Y, Zhang J, Guan Y, Xu D. Analysis and design of a two-phase series capacitor dual-path hybrid DC-DC converter. IEEE Transactions on Power Electronics. 2022;**37**(8):9492-9502

[13] Hua L, Luo S. Design considerations for small signal modeling of DC-DC converters using inductor DCR current sensing under time constants mismatch conditions. In: IEEE Power Electronics Specialists Conf. Vol. 2007. Orlando, FL; 2007. pp. 2182-2188

[14] Gong M, Zhang X, Raychowdhury A. Non-isolated 48V-to- 1V heterogeneous integrated voltage converters for high performance computing in data centers. In: IEEE 63rd International Midwest Symposium on Circuits and Systems (MWSCAS). Springfield, MA, USA: MWSCAS; 2020. pp. 411-414

[15] Bui DV, Cha H, Nguyen VC. Asymmetrical PWM series-capacitor high-conversion-ratio DC–DC converter. IEEE Transactions on Power Electronics. 2021;**36**(8):8628-8633

[16] Rmila SA, Chen Z. Eliminating deadtime mismatch due to inserting storage capacitor of a GaN-based two-phase series-capacitor buck DC-DC converter. e-Prime—Advances in Electrical Engineering. Electronics and Energy. 2022;**2**:100075

Section 2

Microwave and Radio Frequency

Chapter 4

Overview and Advancements in Electric Vehicle WPT Systems Architecture

Victor Oluwaferanmi Adewuyi

Abstract

Wireless Power Transfer (WPT) system is a rapidly evolving technology with vast potentials in consumer electronics, electric vehicles, biomedicals and smart grid applications such as Vehicle to Grid (V2G). Hence, this article is devoted to present an overview of recent progress in WPT with specific interest in magnetic resonance WPT and its system architectures such as compensation topologies, inputs and outputs, as well as coil structure. The strengths, drawbacks and applications of the basic compensations (SS, SP, PS, PP) and hybrid compensations (LCC and LCL) were presented and compared. Although primary parallel compensations perform well at low mutual inductance, they are rarely used due to large impedance and dependence of coefficient coupling on the load. Hence, the need for extra-compensations forming hybrid topologies, such as LCC, LCL, which usually choice topologies for dynamic WPT application or V2G application.

Keywords: compensation topologies, electric vehicle, vehicle to home, magnetic resonance, inductance

1. Introduction

In recent years, research trends on Electric Vehicles (EVs) have gained accentuating attention with focus on advancing the technologies for wireless power transfer (WPT) and applications such as Vehicle to Grid (V2G). Since growth in WPT in electric vehicles creates new possibilities in the V2G, it is hard to talk about the future of Vehicle to Grid (V2G) without including an efficient wireless power transfer system. In designing an effective V2G that integrates electric vehicles into the power grids, a bi-directional WPT system is required such as allows energy flow between the vehicle and the electrical grid [1, 2]. One of the major goals of EV WPT research is the development of alternative systems for powering battery reliant devices to limit the bottlenecks and issues related with the use of wired charging for batteries. Some of the outstanding benefits of WPT system includes improving portability and user-friendliness of devices; by eliminating wires; devices can now be developed with more flexibility and limited environmental design constraints as in the case of medical implantation and underwater *witricity*.

The philosophy of V2G thrives on the premise that EVs are parked for most of the day and the energy stored in their batteries can be utilized during this time provided that their initial state of charge (SoC) is restored before vehicle utilization by the user [2–4]. Although the traditional V2G is designed for wired charging, the system can be easily extended to WPT provided that all the unidirectional stages are replaced with bidirectional stages such that power can flow in both directions [1]. In a situation where the EV charging station is a domestic or residence, the V2G is known as Vehicle to Home (V2H) [5, 6], and in this case, the customer can either use the energy stored in the battery to power its own domestic loads or inject it into the grid if the loads are low-power [7–9]. A design procedure for a bi-directional WPT Vehicle to Home (BWV2H) application was presented, ensuring necessary constraints such as the rules for low voltage utilities and the standard SAE J2954 for EV WPT were satisfied [3]. It is expected that the widespread adoption of BWV2H technology will aid demand response, the smart grid transition, and the seamless integration of renewable energy sources. The V2H technique also helps to minimize the overall power cost by stabilizing or reducing the fluctuating power demand of home loads, in addition to compensation of users for the distribution system's service [3]. Some of the prospects of the study is the optimization of the system to allow load balancing, filling and peak shaving [10, 11]. A mathematical model for the estimation of the charge and discharge efficiency in an EV bidirectional ICPT wireless charger was presented [12]; the WPT system was characterized and validated experimentally using a 3.7 kW prototype. It was suggested that V2G systems need improvement in operational mechanism, compliance and control.

In the past few decades, there has been interesting progress in the wireless power transfer field, which has in turn triggered development and innovation in fields of consumer electronics, electric vehicles, biomedical etc. The transfer of 60 watts of power over a distance of 2 meters with an efficiency of 40% in 2007 by MIT researcher [13] opened up a new landscape of development in wireless power transfer. Other feats include the powering of 14 watts compact fluorescent lamp wirelessly in 2011 using a circular domino with repeater arrangement by researchers from Hong Kong Polytechnic University [14] and the transfer of 5 kilowatts of power over a gap of 20 cm at an efficiency of 90%, by researchers from Utah State University in 2012 [15]. Also in 2012, electric buses were deployed and charging pads were installed at bus stops for charging electric buses by an Uttah-based company, Wireless Advanced Vehicle Electrification (WAVE) [16]. The development and introduction of Online Electric Vehicle (OLEV) into public transit network in 2013, by the Korea Advanced Institute of Science and Technology (KAIST). The OLEV is an electric vehicle that can be charged both in-motion and in-stationary, and the charging system allows for a one-third reduction in the battery compared with that of a regular electric vehicle [17].

The analysis of the performance indices of wireless power systems [18–21], is an integral part of WPT research usually done using equivalent circuit, coupled mode theory or two port networks as modelling technique with a goal of improving the system performance [22]. In order to achieve a stable and functional WPT system for robots, an analysis of the WPT coupling mechanism, coupling coils and compensation topology was presented and extended into investigation of other means of enhancing coupling performance such as Multi-layer coil structure [23]. The use of circuit theory and mutual inductance in modelling various components of the system and solving for performance indices is also a common approach [18–21]. Using equivalent circuits and Neumann's formula [24], the relationship between maximum efficiency and air gap length in magnetic resonant coupling was analyzed and the conditions required to achieve maximum efficiency for a given air gap were proposed. The air gap length was confirmed to be related to the radius and number of turns of the coils. Maximum efficiencies are achieved at various air gap lengths through mutual inductance, characteristic impedance, internal resistance and resonance frequency by setting the optimized characteristic impedances in each case. To boost efficiency of WPT, Theodoropoulos et al. devised a control algorithm for load balancing in wireless EV charging [25]. A solution to the problem of pad misalignment was proposed by Zhao *et al.*, using a combination of the different resonant networks [26].

This work presents an overview and advancements in wireless power transfer, with specific interest in application of magnetic resonant coupling to the charging of electric vehicle. In the foregoing sections, the overview of the WPT system classification is presented; the system architectures were presented with focus on compensation topologies and the coil design aspects and charging standards.

2. WPT system classifications

WPT systems are generally classified using system attributes such as coupling strength, coupling type, transmission direction and transmission distance.; and when considering transmission distance, it can be near-field (non-radiative) or far-field (radiative) based on proximity (radiative). The requirements of the desired application largely determine which type of WPT is recommended (**Figure 1**).

2.1 Coupling strength

When considering coupling strength, WPT can be loosely coupled WPT (LC-WPT) or strongly coupled WPT (SC-WPT). Loosely coupled WPTs primarily consists of inductors and resistors, and they are used in applications where the coupling coefficient is low. On the other hand, strongly coupled WPT achieves improved efficiency through resonance and it finds wider range of usage in consumer applications. However, both LC-WPT and SC-WPT operate at the near-field region of the transmitting antenna. In EV WPT, SC-WPT is considered preferable due to optimal possibility of power transfer when the transmitter and receiver are synchronized to

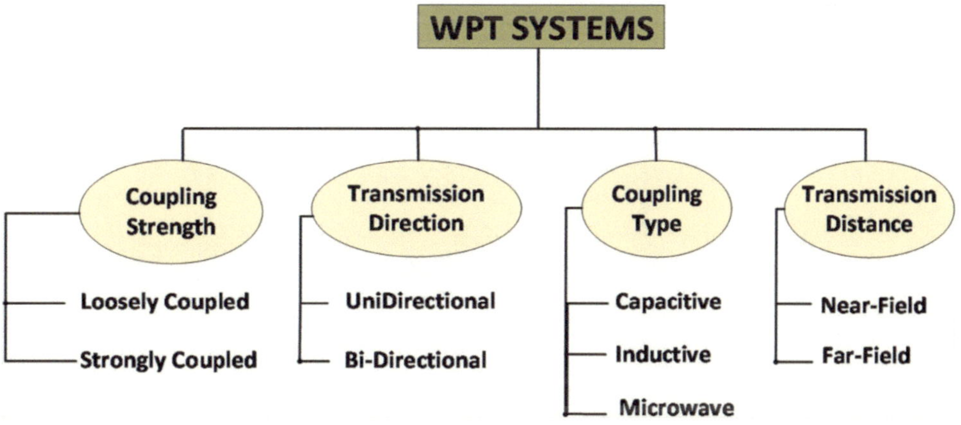

Figure 1.
Classification of wireless power transfer systems.

operate at resonance, hence eliminating reactive power which otherwise would increase system losses. The inductive wireless power transfer, which happens to be the most popular is based on Ampere's Law and Faraday's law.

When current flows through an inductor with N turns, a magnetic flux ϕ is produced. The voltage produced in the coil is proportional to the number of turns N and the time rate of change of the magnetic flux ϕ, according to Faraday's law [27] (**Figure 2**).

$$v = N\frac{d\phi}{dt} = N\frac{d\phi}{di}\frac{di}{dt} = L\frac{di}{dt} \tag{1}$$

Where L is the self-inductance of the coil, given as $L = N{}^{d\phi}/_{di}$. When another coil L2 with N2 turns is brought in close proximity to the current carrying coil L1, such that there is magnetic flux linking, the voltage in coil L1 and L2 are respectively given as;

$$v_1 = N_1\frac{d\phi_1}{di_1}\frac{di_1}{dt} = L_1\frac{di_1}{dt} \tag{2}$$

$$v_2 = N_2\frac{d\phi_{12}}{di_1}\frac{di_1}{dt} = M_{21}\frac{di_1}{dt} \tag{3}$$

Where $\phi_1 = \phi_{11} + \phi_{12}$ such that ϕ_{11} is the flux linkage on coil L1 and ϕ_{12} is the magnetic flux linkage of coil L1 on L2. The mutual inductance is the tendency of one inductor to induce voltage in a nearby inductor, given as;

$$M_{21} = M_{12} = M = N_2\frac{d\phi_{12}}{di} \tag{4}$$

$$M_{21} = k\sqrt{L_1 L_2} \tag{5}$$

Where the coupling coefficient, $0 \leq k \leq 1$, is a measurement of how magnetically connected two coils are.

2.2 Coupling type

When considering coupling, WPT can be capacitive coupling (C-WPT), inductive coupling or inductive power transfer (IPT), magnetic resonance coupling (MRC), and

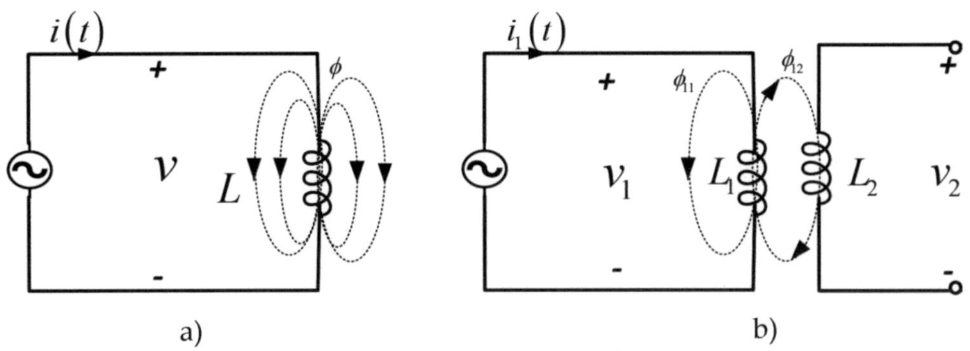

Figure 2.
(a) Self Inductance, (b) Mutual Inductance.

microwave coupling. Inductive coupling involves transfer of energy from the magnetic fields of an inductor on the transmitting end to another inductor on the receiving end. Although it has the ability of achieving a high efficiency (>90%) [28] and it requires minimal area for magnetic influence, a main drawback is the limitation in the distance of transfer between the transmitter and receiver. The efficiency reduces drastically as separation distance increases [29], and the operating frequency typically ranges from 10 kHz to MHz [30].

Resonant Inductive Coupling is an approach used for reducing system impedances of inductive coupling WPT systems by compensating the primary and secondary coils with capacitors, leading to higher power transfer efficiency through magnetic resonance [13]. It is also found that the coil parasitic inductance can be used for compensation, provided the operating frequency is significantly high enough for the coil's self-resonant frequency [31, 32]. The differences between IPT and MRC is the presence of compensation capacitors in MRC for tuning the secondary and the primary coils to a resonant frequency, whereas, IPT WPT has no resonant circuit involved. Also, MRC is preferred for low- or medium-powered WPT applications while IPT WPT is preferred for high-voltage WPT applications [1].

Capacitive coupling is unique in its coupling approach as it engages capacitance effect in transmitting power from the source to the receiver such that the transmitter's set of electrode couples with the set of electrodes at the receiving end [33]. This makes it considerably suitable for low power lightweight application that is in few kW. However, safety concerns increase as the power rises and large electric fields are transmitted.

2.3 Transmission distance

The near field WPT includes loosely coupled, strongly coupled, inductive coupled, capacitive coupled and any other coupling whose transmission distance is in the near-field region. Microwave coupled WPTs operate in the far-field zone. Due to the possibility of transmission distance to meet design requirements, magnetic resonance coupling is gaining momentum in applications such as Electric Vehicle (EV) charging. In near-fields, the power transfer distance is shorter than the operating wavelengths, and examples include the inductive and capacitive coupling. While the efficiency of power transfer is much higher than the far-field techniques and has higher misalignment tolerance, the transfer distance is small and efficiency deteriorates with increasing distance.

Far-field WPT are characterized transfer distance significantly larger than the electromagnetic wavelength and can power devices at tens of thousands of kilometers. Examples include wireless power transfer from solar satellites to ground rectennas. However, a major drawback is the low efficiency and directivity constraints; the misalignment tolerance is too low hence requiring high alignment for successful transference of power. In order to improve the transmission characteristics of wire antennas for wireless power transfer, such as directivity, analysis of dipole transmitter elements for WPT was carried out [34]. Using the half-wave dipole, the current distribution and radiated fields were computed via Method of Moment technique (MoM). The design and analysis of 5, 6, 7, 10, 20 and 30 elements of broadside and endfire arrays at 0.3, 0.4 and 0.5 inter-element spacing. Despite the fact that far-field evanescent resonant-coupling techniques can transmit energy over longer distances than near-field induction methods [35, 36], much research attention is focused on near-field power transmission [17, 37], which has applications in consumer electronics, biomedical devices, electric vehicles, and other areas [20, 38, 39]. This is owing to the fact that when utilizing the radiative approach, the efficiency of wireless power transfer drops dramatically.

2.4 Transmission direction

When considering the direction of energy transmission, WPT can be unidirectional or bidirectional the unidirectional is the commonly used WPT for electric vehicles. However, when the system requires power transfer in both direction as in the case of V2G, a bidirectional system is deployed. The figure below shows the equivalent circuit topology of a bi-directional WPT for V2G, containing three basic elements; the DC voltage source, the inverter, and a compensation network [40] (**Figure 3**).

Both the primary and secondary sides of a bi-directional WPT system can act as transmitters or receivers. Consequently, both DC sources must be able to release and absorb energy. In practice, a rectifier connected to voltage input is frequently used as the primary voltage source, and a battery stack is frequently used as the secondary voltage source [40].

There are also four power electronic switches, which serves as inverter on the primary side, and as rectifier on the secondary side. Compensation is also introduced so as to improve the efficiency caused by low coupling leading to high reactive power.

3. System architectures

When designing the circuit architecture of the WPT system, there are certain design objectives required, the system losses increase with increasing reactive power. To achieve this, the circuit is compensated using capacitors, so as to achieve one or more of the following requirements [17, 41, 42]:

1. **Power Transfer:** It is mostly desired to maximize the power transfer and this is achievable by cancelling leakage inductances on both primary and secondary sides [17, 43–45].

2. **VA rating:** It is usually desired to minimize the VA rating of the power supply by providing regulated reactive power needed for establishing and sustaining the magnetic field [17, 46]. Reactive currents increase semiconductor and

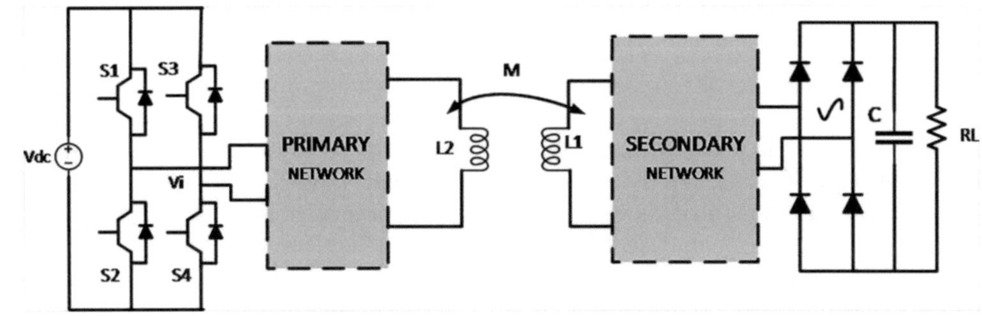

Figure 3.
Circuit Topology of bidirectional WPT system.

conduction losses, mostly in diodes. The primary leakage inductance is cancelled by primary resonance to increase the power factor to near unity.

3. **Phase Angle and Soft Switching**: Since the VA rating on the power supply is determined by the phase angle between the input voltage and the current, the minimum VA rating requires the phase angle to be zero while the switching of MOSFETs requires that phase angle be large than zero. Hence, phase angle is usually selected to be slightly greater than zero to achieve soft-switching, and low enough for acceptable VA rating [42].

4. **Constant Voltage (CV) or Constant Current (CC):** It is sometimes desired to provide Constant Voltage or Constant Current at the output. Usually, the end load for a wireless power transfer system is a battery and it requires constant current while charging at low states (0–85%), and constant voltage for charging state [17, 47]. In some applications, compensation networks are specially designed to keep the output voltage constant or output current constant so as to meet the battery charging requirements [17, 42]. Primary series compensation is necessary for voltage source, while parallel compensation is used for current source drive.

5. **Bifurcation Tolerance:** It desired to improve the Bifurcation tolerance of the system. This is done by designing the compensation network for single zero phase angle, while ensuring system stability for different loading conditions and frequency variation control [17, 41, 42].

6. **Misalignment Tolerance:** Each compensation topology has a varied sensitivity to altering location (misalignment). This necessitates a more complex control mechanism to ensure that the resonance frequency is maintained on both sides.

Table 1 shows a summary of the Requirements and the dependencies. In addition to the aforementioned points, frequency splitting usually occurs in strongly coupled systems, where the electromagnetic coupling between coils are high [22], leading to various adaptive controlling measures to maximize the power transfer; such as frequency tuning [48, 49]; impedance matching [50, 51], and coupling manipulations [52, 53]. One of the two goals of WPT is to attain maximum power transfer, and this occurs when the source impedance and the input impedance are matched. Recent

Requirement	Dependency
Maximum power transfer	Compensation topology
Minimal VA Rating	Primary resonance
Zero phase angle	Soft switching
CV or CC output	Load, compensation, topology
Bifurcation tolerance	Load quality factor, compensation topology and
Misalignment tolerance	Compensation topology
Bidirectional flow	Compensation topology, power control

Table 1.
WPT system requirements for compensation network designs.

WPT research has been focused on strategies for increasing PTE in wireless EV charging systems and improving pick-up power [54].

3.1 Compensation topologies

As mentioned in the last section, compensations are introduced to obtain desired characteristics, when the coupling coefficient reduces to less than 0.3. The WPT pads are loosely coupled and have large leakage inductance and hence prompts the need for a compensation network to suppress this. Compensation circuits are typically required for both TX and RX in order to increase the system's power transmission metrics such as power transfer efficiency (PTE) and power delivered to the load (PDL). PTE and PDL are two crucial characteristics in WPT technology that influence the range of power transmission and interference with other devices [55].

A compensation circuit has several advantages, including lowering the power supply's volt-ampere (VA) rating and more efficiently adjusting the value of current in the supply loop as well as the voltage in the receiving loop. The easiest way to achieve compensation is adding a capacitor to each side of the system, and this produces basic topology or hybrid topology. Factors that influence the selection of the L and C values of the resonant circuit includes the primary or secondary topology, the quality factor and the value of the magnetic coupling coefficient [56].

3.1.1 Basic compensation topology

There are four basic compensation circuits and numerous hybrid compensation circuits in MRC WPT. The four basic compensation systems are Series–Series (SS), Series–Parallel (SP), Parallel–Series (PS) and Parallel–Parallel (PP). The figure below shows the circuit schematic of the four basic compensation networks (**Figure 4**).

Where the impedance for primary series is given as $Z_p = j\omega L_p + R_p + \frac{1}{j\omega C_p}$, secondary series impedance is given as $Z_s = j\omega L_s + R_s + \frac{1}{j\omega C_s}$; and secondary primary impedance is $Z_p = j\omega L_s + R_s + \frac{R_L}{j\omega C_s R_L + 1}$. The table below presents the resonant

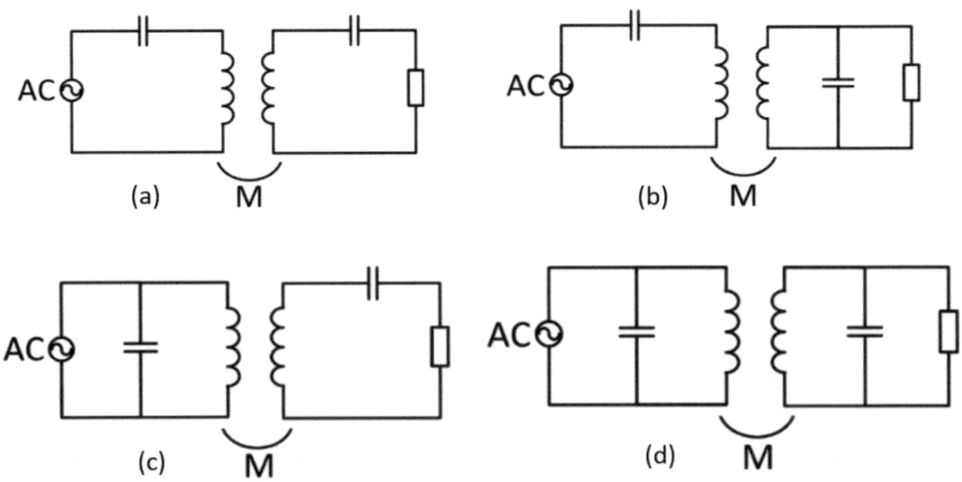

Figure 4.
Circuit Schematic of Basic WPT Topologies (a) SS, (b) SP, (c) PS, (d) PP.

DOI: http://dx.doi.org/10.5772/intechopen.106254

Topologies	C_p [56]	Z_{in} [57]	η [57]
SS	$\frac{1}{\omega_r^2 \cdot L_p}$	$Z_p + \frac{\omega^2 M^2}{Z_s}$	$\frac{\omega^2 M^2 R_L}{\lvert Z_s\rvert^2 R_p + \omega^2 M^2 (R_L + R_s)}$
PS	$\frac{L_p}{\left(\frac{\omega_r^2 \cdot M^2}{R}\right) + \omega_r^2 \cdot L_p^2}$	$\frac{Z_p - \frac{1}{j\omega C_p} + \frac{\omega^2 M^2}{Z_s}}{j\omega C_p \left(Z_p + \frac{\omega^2 M^2}{Z_s}\right)}$	
SP	$\frac{1}{\left(L_p - \frac{M^2}{L_s}\right)\omega_r^2}$	$Z_p + \frac{\omega^2 M^2}{Z_s}$	$\frac{\omega^2 M^2 \frac{R_L}{1+\omega^2 C_s^2 R_L^2}}{\lvert Z_s\rvert^2 R_p + \omega^2 M^2 \left(\frac{R_L}{1+\omega^2 C_s^2 R_L^2} + R_s\right)}$
PP	$\frac{L_p - \frac{M^2}{L_s}}{\left(\frac{M^2 \cdot R}{L_s^2}\right) + \omega_r^2 - \left(L_p - \frac{M^2}{L_s}\right)^2}$	$\frac{Z_p - \frac{1}{j\omega C_p} + \frac{\omega^2 M^2}{Z_s}}{j\omega C_p \left(Z_p + \frac{\omega^2 M^2}{Z_s}\right)}$	

Table 2.
WPT System requirements for compensation network designs.

capacitors, input impedance and efficiency of the basic compensation networks (**Table 2**) [56, 57].

Series–Series (SS) Compensation: In SS, unity power factor occurs at resonant frequency, making the compensation desired where high efficiency and high-power factor (PF ≈ 1) are required. The importance of these two metrics is hinged on how they fluctuate with coupling and load variation [42]. One of the advantages of the SS compensation includes the independence of the value of the primary capacitance on coupling coefficient and load [45, 47]. It finds application in dynamic WPT where the coupling coefficient changes as the device moves; an example is the segmented dynamic WPT charging in EV. Series compensation, in which the compensation capacitors are connected in series along the primary track, is ideally suited for systems with lengthy primary tracks. This permits the track voltages to be kept within acceptable levels. Because concentrated windings are often high-current systems, parallel compensation is better suited for them [56]. A major drawback however is the light-load condition that occurs when the receiver is not present, thereby creating a zero-impedance at the primary resonance frequency. Hence, the voltage transferred to the load under this condition is very high. The only thing that limits the current in this situation are the parasitic impedances of the coil and capacitor.

1. Series–Parallel (SP) Compensation: In SP, regardless of the load, there are some impedances that are transferred to the primary and at resonance, there would still be a short circuit if the load is absent which facilitates the need for a current limiting control. Also, variations in the mutual inductance leads to changes in the power factor and system dynamics, hence making power factor more challenging [42, 45]. Also, as the mutual inductance changes, the resonant frequency and consequently, the value of the capacitance are changes. The SS and SP compensation are most commonly utilized in actual applications and implementation because they have the highest efficiency, with efficiencies exceeding 93 percent and approaching 97 percent at various transmission power levels [56, 58, 59].

2. Parallel–Series (PS) Compensation: The reflected impedance to the primary is the same for the series–series and parallel–series configurations. At relatively low mutual inductances and relatively large range of changes in load and mutual inductance, the efficiency and power factor are high [60, 61]. Also, since the system requires current source input, to avoid instantaneous changes in voltage, the P-S is compensated by adding an inductor to create an LCL resonant task. A

study shows that variation in coupling influences the compensation capacitance required in secondary parallel circuits [62]. It is also possible to combine the characteristics of SS and PS to achieve SPS constant output power without adjusting the power supply [11, 63]. Although PS allows for soft switching in semiconductors [64], compensations on the primary side are rarely employed due to high impedance, calculation complexity, and coupling coefficient dependence on load, among other factors [65]. In PS and PP, the increased input impedance necessitates a high driving voltage to transfer sufficient power [66].

3. Parallel–Parallel (PP) Combination: Since series compensations requires higher voltage and current than parallel compensation, a combination of both is used to obtain the required capacitance needed to achieve the desired voltage and current ratings. Also, a combination of series–parallel–series topology is proposed to suppress the effect of misalignment [63]. In general, analysis of the four basic compensation circuit topologies shows that the maximum power transfer capability is attained by reducing the reactive power flow through varying the operating frequency; however, control complexity caused by frequency bifurcation occurs and triggers system instability [67].

3.1.2 Hybrid compensation topologies

In order to achieve further improvement such as constant current/voltage, higher efficiency, designs of hybrids of capacitors and inductors topology have been explored for the MR WPT circuit, such as LCL, LCC, LCCL etc. The LCL and LCC are considered briefly below:

1. LCL Compensation: This is formed by adding an inductor to parallel resonance network [44] and it has the ability of overcoming the problems associated with series and parallel compensations. Using an LCL compensation [1], a phase-angle control method was introduced to effectively regulate the direction and magnitude of power flow in a WPT device, and a mathematic model of a bidirectional power flow was presented. As a strength, the source current of LCL can be easily controlled for variations in coupling coefficient and load conditions using a full bridge converter with minimized VA rating. The secondary side can be a parallel compensation, series compensation or hybrid. Common secondary combination of LCL primary compensation includes, LSL-S, LCL-P, LCL-LCL. In order to achieve constant current and constant voltage in the transmitting coil, a dynamic LCL-S/LCL topology was proposed [68]. Parallel compensation is popular because it offers flexibility to variations in load [69]. However, the reflected impedance on the primary has both real and imaginary components of the load, which contributes to reactive power. To ensure continuity of conduction through the rectifier, a large dc inductor is required, which increases the cost and loss in the system. Comparing LCL-S with LC/S, they are identical in structure but have different tuning methods. However, the LC provides a better load independent voltage output and strong capability of high voltage suppression than LCL-S [70]. Also, the LC compensation network with LCL in the primary side performs like a current source at the resonant frequency irrespective of the coupling and load condition because the current in the primary side coil is controlled by the high frequency square wave voltage from the power converter. This makes LCL adopted for many designs [15, 71, 72],

LCLC compensation is employed in cancelling the nonlinear effect of the rectifier diodes, to achieve an exactly unit power factor at predetermined load condition. Both LCL and LCLC achieve remarkable improvement in efficiency as compared with traditional LC parallel structure.

2. LCC Compensation: The LCC is formed by adding an inductor in series and a capacitor in series to a parallel resonant circuit. Several researches have been conducted on the double sided LCC [73–77]. As a strength, it can be used to compensate the power factor at the secondary side to achieve unity power factor. It is also independent of coupling coefficient and load conditions while ensuring Zero Volt Switching (ZVS) for MOSFETs [78]. In general, researchers favor double sided LCC because of the high misalignment tolerance, load independence, reduction of current stress in the inverter, higher efficiency [73]. A comparative analysis between S-S and LCC-LCC in EV WPT shows that LCC-LCC outperforms S-S topology in efficiency stability with respect to variations in self inductance due to lateral displacement of receiving and transmitting coils [76]. But when compared to the LCCL compensation topology, LCCL has a higher efficiency and a higher power transfer level, as well as a larger coupling coefficient. Compared with LC-S, LC-S experiences a 2.5% improvement in efficiency [79].

3.2 Variation in number of transmitters, receivers and stages

Based on the number of transmitters or receiving elements, the WPT system can be classified as single-input single-output (SISO), multiple-input single-output (MISO), single-input multiple-output (MISO) or multiple-input multiple-output (MIMO).

3.2.1 SISO, SIMO, MISO and MIMO

SISO and MISO: SISO WPT is a basic and simple WPT System prototype in which the distance and orientation of the coils have a significant impact on the reflected load. In MISO WPT, cross coupling occurs between the transmitters which increases the effects of the load on the transfer function and system efficiency. Two scenarios of system node coupling are Increased Density and Constant Density. In increased density scenario, the number of receivers increases within a constrained region as well as the cross couplings between receivers, while the mutual coupling between receivers and transmitters remains constant. This results in reduction in point to point efficiency as number of nodes increases. In constant density scenario, the region and the number of receivers are increased, mutual coupling is reduced while cross coupling between transmitters remains constant. The power efficiency and output power vary as number of receivers are changes.

SIMO and MIMO: In SIMO WPT, the systems performance is influenced not only by the mutual coupling between transmitter and receivers, but also by the cross couplings between receivers as well. MIMO WPT is popular for wireless sensors and other electromagnetic radiation WPT applications [80–82]. Drawbacks in designing efficient MIMO include determining exact resonant frequency. An algorithm was proposed [83] for isolating the desired receiver and maximizing the power while limiting power to unintended receivers. Notable advantages of SIMO over MISO include greater power transmission efficiency (PTE) and lower magnetic field strength requirements, making it safer for human health [79, 84]. As a result, the MISO WPT system is a strong contender for biomedical implants, general electronics, and dynamic electric car charging.

3.2.2 Variation in transmission stages

The WPT system can also be designed using more than two coils, such as the 3-, 4-coil etc. The two-coil WPT is the basic SISO prototype, consisting of two electromagnetic subsystems with the same natural resonant frequency. Using the resonance frequency as a key parameter, mathematical expressions of optimal coupling coefficients of 2-coil WPT were examined and the system efficiency was analyzed with respect to the air gap values for various [85]. It was shown that the maximum PTE always occurs at the resonance frequency and reduces with decline in coupling coefficient. The efficiency values were calculated using three kinds of softwares in order to determine the difference between the software outputs. The results show that the equivalent circuit analysis by means of numerical computing is best suitable for determining the voltage and current waveforms. It was suggested that minimum distance between coils should be used to keep the energy transfer at an optimum, and this can be solved by optimizing the relation between the quality factor and frequency. A 4-coil system, which consists of the transmitter (source coil and a sending coil) and receiver can be conventional or unconventional (asymmetric). In conventional 4-coil [86–88], the input impedance of the system can be adjusted by mounting and coupling the source coil with the sending coil. Similarly, the equivalent load resistance can be adjusted to match the load condition by mounting and coupling the load coil with the receiving coil. The asymmetric 4-coil system, proposed by Moon [89], is made up of primary side (source coil and two transmitter coils) and secondary side (receiver or load coil). When compared to conventional 4-coils, the asymmetric 4-coils have a higher PTE and a longer range between the source and load coils; this is owing to improved augmentation of the apparent coupling coefficient.

3.3 Static and dynamic charging

Static Charging finds application in designs where the electrical loads are required to be stationary during the wireless powering. Examples include consumer electronics [90], biomedical devices [91, 92] and electric vehicles [16, 72, 93]. In a static WPT, the source transmitting coil is excited with a high frequency electric field and transmits magnetic field (B) to the receiving coil through the air gap. The electromagnetic field, landing on the receiving coil induces an electric current in the coil capable of transmitting several watts of power across the air gap. One of the common applications of static WPT today is the electric vehicle WPT charging stations where charging cables, extensions and sockets are replaced with coils and also aid the easy realization of complete design and implementation of autonomous vehicle operations. Some of the design goals for static EV WPT include [79]; maximizing the power transfer efficiency (PTE) for a given cost and specification; increasing the magnetic coupling to increase induced voltage; compactness in size; and managing fluctuations in resonant frequency and coupling coefficient caused by pick-up position misalignment and air-gap variation. In stationary EV WPT, the utility power is converted from low frequency (LF) to high frequency (HF) by high frequency inverter. This generated HF electromagnetic field energizes and transfers power to the receiving resonator. Generally, wireless power converters can be direct or indirect power converters. Although the transmitter track is easier to control, some of its major drawbacks include safety need to suppress harmful electromagnetic emissions; also, the need to distribute compensation capacitor to minimize the large inductance is costly, low coupling coefficient and others.

Dynamic wireless power transfer (DWPT) allows for mobility of load during the transfer of power with vast applications in electric vehicles as well as bio-medicals, where ingested sensors can continue operating and charging while customer is breathing or moving around [94]. In EV, charging of the vehicle while in motion promises to help improve the requirements for driving range and battery charging of EVs. Dynamic WPT can be classified based on the transmitter array design; either a single transmitter track and or segmented transmitter coil array. The single transmitter usually consists of a long transmitter track connected to the power source, while the receiver is smaller in comparison. Segmented coil array usually consists of multiple coils of transmitting resonators connected to a high frequency power source. While the dynamic suppresses field exposure and the need for distribution of compensation capacitors, it also presents peculiar challenges some of which include; need for right optimization of transmitter coil design parameters, such as setting the appropriate distance without compromising the mutual inductance and design cost. The dynamic EV wireless charging already finds application in electric buses and trams at low speeds in urban areas [94]. Despite the prospects of the DWPT, issues such as the problem of accurately forecasting and responding to private EV's dynamic charging demands is still a major drawback since the routes and speed are largely unpredictable. Unlike plug-in and static WPT that charges for hours at low power ratings, DWPT is expected to take few seconds or few minutes. Another challenge is the need to keep track of receiver position and regulating power supply appropriately as the load navigates the array. Some enquiries relevant to the advancement of this research direction include finding impact of inter-coil mutual inductance on the overall power transfer efficiency, influence of inter-element distance on coil efficiency and how this can be optimized. Examples of the track based dynamic WPT includes [95, 96]; the online electric vehicle (OLEV), designed by Korean Advanced Institute of Science and Technology (KAIST). The OLEV has 5 to 60 m long rails powering pickup modules of 80 cm in length. Another notable example is the track-based dynamic wireless power transfer system of UC Berkeley.

4. Coil designs

The coil architecture is one of the critical areas of consideration in the design of an optimal WPT system because of its influence on the minimum efficiency requirements, misalignment tolerance, cost, volume, weight and other performance benchmarks as assessed and presented using analytical and numerical methods [37, 97–99]. For instance, the shape of the coil affects transmission efficiency and studies show that different shapes (spiral, square, circular, solenoidal) all have different efficiencies [100–102]. A comparison between the solenoidal coil and spiral coil was presented using experimental analysis [103]. The impact of each parameter of a circular coil structure was investigated in order to determine the limit of design parameters in improving the coil performance, and determination of misalignment performance which leads to exploration of alternative structures for coils, such as the double D coils (DD-coils) and double D quadrature coil (DDQ coils) [104, 105, 106–108]; and XPAD [109]. The use of circular coil of Litz wire was proposed for transferring 1 kW power through an air gap distance of 300 mm at 100 kHz, and achieved an efficiency greater than 80% [110]. In addition to the coil shape, the structural composition of the coil also affects the system performance. For instance, it is proposed that an efficiency of 40% can be achieved at 0.5 cm transfer distance [111] using a coil of high Q planar-Litz. To increase the inductance, Mizuno *et al.* proposed a magnet plated copper wire with a magnetic thin layer coated around the circumference to boost inductance; and the resistance due to proximity effect is lowered because eddy current loss is minimized [112]. Unlike in EV WPT system, space limitation is a major design constraint in some applications such as medical implants, thereby furthering the need to optimize the coil design [113]. Some of the coil parameters that affect the system performance include; ratio of coil diameter to air-gap (R:L); coil geometry, Q-factor etc. [114]. A study shows that the transmission efficiency exceeds 80% when air-gap (L) is lesser than half of coil diameter (R), i.e. L/R < 0.5, and the transmission efficiency reaches 90% when L/R < 0.25 [114].

4.1 Coupling pads

The coupler is considered the most important part of the WPT system [17], and it consists of the transmitter and receiver coils separated by a magnetic gap. When designing coupling coils, the desired attributes include: high coefficient of coupling (k); high quality factor (Q); and high misalignment tolerance [62, 115].

High Q: Designing the inductors to have high self-inductance and low series resistance at high frequency helps improve the Q. However, considering the standard by SAE J2594/1, the maximum operating frequency is limited to approximately 85+/− 3.7 kHz. Hence, increasing the quality factor means reducing the resistance of the coil.

$$Q = \sqrt{Q_1 Q_2} \tag{6}$$

Where $Q_{1,2} = {}^{\omega L_{1,2}}/_{R_{1,2}}$ and $L \, \alpha \frac{\mu N^2 A}{l}$.

The Coil inductance is directly proportional to the square of the number of turns, while the ESR is directly proportional to the no of turns. Although, increasing number of turns also increases ESR but it increases inductance more. Hence the coil Q can be increased by increasing the number of turns [116, 117]. Hence to find an optimum self-inductance for the coil, a balance must be found between the wire diameter and the number of turns. Alternatively, ferrite bars are proposedly used on the coil to increase self-inductance by guiding the flux so that leakage flux can be reduced and high coefficient of coupling can be realized [17]. The coil ESR consists of DC and AC resistances. The DC resistance can be reduced by increasing the area of the conductor, while the use of Litz wire is adopted to reduce the AC resistance [118]. The product of kQ, which affects the efficiency of the coupled inductors in the system, is determined by the geometry, the core material as well as the magnetic gap [116, 117, 119].

High Coefficient of Coupling (k): In order to improve the coefficient of coupling and maintain high quality factor, several coil structures such as Circular Pads, Flux Pads, DD Coils, Multi-Coil Polarizer have been proposed [30, 117, 120].

1. Circular Pads – Several reports on the design and optimization of circular pads have been reported in the literature [116, 117]. Using Pareto for optimization, guidelines for optimizing the design of circular coils with respect to area-related power density and efficiency are discussed [116, 117]. It was proposed that the diameters of the coils should be 600-800 mm to keep the coupling coefficient in the range of 0.15–0.2, when the magnetic gap is 150–200 mm [118]. A good

assessment of WPT for automotive applications was also provided, as well as techniques for sizing circular pad and performance metrics in EV [30].

2. Flux Pipe – The use of rectangular bars of ferrite with a coil wounded along the length is also proposed, and report shows that it exhibits good lateral misalignment tolerance and further guides the flux to terminate at the ends of the coils [104, 120]. Although it has the merit of increasing horizontal misalignment tolerance and providing fundamental flux path that is half the receiver pad's length, the main drawback which causes loss of Q and low efficiency is that it is solenoidal. This results from the aluminum shielding on the solenoidal coil, which causes loss of quality factor when it is close to one side of the coil and has flux interception [120].

3. DD Coils: To combined the advantage of flux pipe and circular pads, the use of polarized single-sided flux coupler was proposed; where the middle portion of the DD coil, similar to the flux pipe, are connected magnetically in series [120]. And it offers a single sided flux path; higher coefficient of coupling resulting from flux path height proportional to half of the pad length; lower losses in aluminum shielding; low leakage flux from back of the coil and improvement of no-load Q [121].

4. Multi-Coil Polarized Coupler: this includes the use of couplers with different coils either on secondary or primary side [120, 121], usually derived from DD coil and circular coils. The advantages include; tolerance to misalignment and minor variations in magnetic gap spacing. Examples include the DDQ Coil, Bipolar Pad, Tripolar Pad etc.

4.2 Coil performance evaluation

Based on ANSYS simulation, the performance of different coil structures were evaluated and the relative benefits of the geometry was compared [17, 116, 117]. With respect to coefficient of coupling and misalignment tolerance, Bipolar Pads demonstrates the highest performance but with a drawback of decrease in coefficient of coupling with addition of aluminum shield. With respect to shielding, circular coil performs better because its coefficient of coupling is not affected much based on the one sidedness of the flux pattern. Hence, bipolar pad and circular pads are the most popular amongst the coupler [116, 117].

The aforementioned improvement procedures provide sub-optimum outcomes in some situations because they search for a single objective while keeping the other parameters constant. Whereas, the objectives are dependent on multiple parameters, and optimizing one parameter may inadvertently affect another parameter. This facilitated the use of evolution algorithm, such as Genetic algorithms to evaluate the coil performance with different parameter combinations [122, 123], and formulation of procedures to handle multi-objective optimization of benchmarks such as coil to coil efficiency [124–126], figures of merit, coupling and Quality Factor as well as cost, weight and volume [125–128].

5. International charging standards

In order to achieve compatibility across different devices, there are standards of regulation such as the Qi Standard, Alliance for Wireless Power (A4WP), SAE, and International Electrotechnical (IEC) standards.

1. The Qi Standard: The Wireless Power Consortium developed the Qi standard, which is suitable for electrical power transfer over distances of up to 40 mm. It finds application mostly in inductive WPT and comprises of two fundamental elements; the Base Stations and the Mobile devices. The base stations, which is the power transmitter provides inductive power for wireless transmission while the Mobile devices are the end loads that consume the wireless power. Qi wireless chargers are classified as low power or medium power depending on their power delivery range. Low-power chargers are those that can give up to 5 watts at a frequency of 110–205 kHz, and most consumer electronics such as cell phones, music players, and Bluetooth earpieces, fall within this category. Chargers in the medium power category can offer up to 120 watts at a frequency of 80–300 kHz. A variety of mobile phones, such as the iPhone 8, iPhone 8 plus, and iPhone X, adopted the Qi certification. Moreover, thousands of Qi wireless charging stations can be found in public places such as hotels, restaurants, coffee shops, pubs, and public transportation. The Qi standard also includes two positioning types, guided and unguided positioning of the mobile device relative to the base station, to ensure enough coupling for effective transmission. Alliance for Wireless Power (A4WP): The A4WP was established to develop and maintain standards for a type of wireless power that allows for more spatial freedom than previously available standards. The A4WP's magnetic field is spread out across a large area, making device location less important and allowing a single power TX to charge multiple devices at once (SIMO). A4WP, on the other hand, supports Z-axis charging, allowing the device to be detached from the charger [79]. Similarly, to Qi standard, A4WP has two main elements; the Power TX unit that is responsible for power transmission and the Power receiving unit.

2. SAE Standard: SAE International, a multinational organization dedicated to serving as the engineering profession's ultimate knowledge source, ratified the TIR J2954 standard for plug-in hybrid EV wireless charging. All light-duty vehicle systems use the 85 kHz (81.39–90 kHz) frequency band established by SAE TIR J2954. WPT levels are also divided into four PH/EV classes: 3.7 kW (WPT 1), 7.7 kW (WPT 2), 11 kW (WPT 3), 22 kW (WPT 4); and higher power levels may be added in future editions [129]. Many wireless vendors and companies, such as Qualcomm, WiTricity, Evatran, and others, are operating wireless charging using J2954 [79].

3. IEC: Some of the standards published by the IEC includes, IEC 61980-1 in 2015, which addresses general requirements for EV WPT systems, including efficiency, electrical safety, electromagnetic compatibility, electromagnetic field (EMF) protection, as well as general background and definitions. The second part of the series, IEC 61980-2, contains communication needs between electric road cars and WPT systems, as well as some background information and definitions. IEC 61980-3, the third part of the series, will contain special specifications for EV magnetic field WPT systems [79].

6. Conclusion

WPT is a rapidly developing technology with enormous possibilities. This article provided an overview of WPT advancements, with a focus on magnetic resonance

WPT and its system architectures; compensation topologies, inputs and outputs, and coil structure. Basic compensations (SS, SP, PS, PP) and hybrid compensations (LCC and LCL) are given and contrasted in terms of their strengths, limitations, and applications. Primary parallel compensations operate well at low mutual inductance, but they are rarely employed due to high impedance and coefficient coupling reliance on the load. As a result, extra-compensations are required, resulting in hybrid topologies such as LCC and LCL, which are commonly used for dynamic WPT or V2G applications. Considerable attention will be paid in future to novel ways of achieving the improved design objectives.

Author details

Victor Oluwaferanmi Adewuyi
Information and Communications Engineering, Changchun University of Science and Technology, Changchun, China

*Address all correspondence to: feranmeadewuyi@gmail.com

References

[1] Shi ZH et al. Modeling and experimental verification of bidirectional wireless power transfer. IEEE Transactions on Applied Superconductivity. 2019;**29**(2):1-5

[2] Monteiro V, Pinto J, Afonso JL. Operation modes for the electric vehicle in smart grids and smart homes: Present and proposed modes. IEEE Transactions on Vehicular Technology. 2015;**65**(3): 1007-1020

[3] Bertoluzzo M, Giacomuzzi S, Kumar A. Design of a bidirectional wireless power transfer system for vehicle-to-home applications. Vehicles. 2021;**3**(3):406-425

[4] Buja G, Bertoluzzo M, Fontana C. Reactive power compensation capabilities of V2G-enabled electric vehicles. IEEE Transactions on Power Electronics. 2017;**32**(12):9447-9459

[5] Dong D et al. Grid-interface bidirectional converter for residential DC distribution systems—Part one: High-density two-stage topology. IEEE Transactions on Power Electronics. 2012; **28**(4):1655-1666

[6] Monteiro V et al. Improved vehicle-to-home (iV2H) operation mode: Experimental analysis of the electric vehicle as off-line UPS. IEEE Transactions on Smart Grid. 2016;**8**(6):2702-2711

[7] Tan T et al. A bidirectional wireless power transfer system control strategy independent of real-time wireless communication. IEEE Transactions on Industry Applications. 2019;**56**(2): 1587-1598

[8] Nguyen BX et al. An efficiency optimization scheme for bidirectional inductive power transfer systems. IEEE Transactions on Power Electronics. 2014;**30**(11):6310-6319

[9] Wang L, Madawala UK, Wong M-C. A wireless vehicle-to-grid-to-home power interface with an adaptive DC link. IEEE Journal of Emerging Selected Topics in Power Electronics. 2020;**9**(2): 2373-2383

[10] Kuramoto S, Akatsu K. Basic experiment on high power transmission at 13.56 mhz wireless power transfer for electric vehicle. In: IEEE Vehicle Power and Propulsion Conference (VPPC). IEEE. 2019

[11] Li S, Mi CC. Wireless power transfer for electric vehicle applications. IEEE Journal of Emerging and Selected Topics in Power Electronics. 2015:4-14

[12] Triviño A, Gonzalez-Gonzalez JM, Aguado JA. Theoretical analysis of the efficiency of a V2G wireless charger for electric vehicles. Transactions on Environment Electrical Engineering. 2018;**3**(1):9-14

[13] Kurs A et al. Wireless power transfer via strongly coupled magnetic resonances. Science. 2007;**317**(5834): 83-86

[14] Zhong W, Lee CK, Hui SR. General analysis on the use of Tesla's resonators in domino forms for wireless power transfer. IEEE Transactions on Industrial Electronics. 2011;**60**(1):261-270

[15] Wu HH et al. A high efficiency 5 kW inductive charger for EVs using dual side control. IEEE Transactions on Industrial Informatics. 2012;**8**(3):585-595

[16] Wang X et al. Electric vehicle charging station placement for urban public bus systems. IEEE Transactions

on Intelligent Transportation Systems. 2016;**18**(1):128-139

[17] Patil D et al. Wireless power transfer for vehicular applications: Overview and challenges. IEEE Transactions on Transportation Electrification. 2017; **4**(1):3-37

[18] Triviño-Cabrera A, Aguado JA, editors. Emerging Capabilities and Applications of Wireless Power Transfer. IGI Global; 2018

[19] Kumar A, Mirabbasi S, Chiao M. Resonance-based wireless power delivery for implantable devices. In: IEEE Biomedical Circuits and Systems Conference. IEEE. 2009

[20] Lee B, Kiani M, Ghovanloo M. A triple-loop inductive power transmission system for biomedical applications. IEEE Transactions on Biomedical Circuits Systems. 2015;**10**(1):138-148

[21] Kiani M, Ghovanloo M. Pulse delay modulation (PDM) a new wideband data transmission method to implantable medical devices in presence of a power link. In: IEEE Biomedical Circuits and Systems Conference (BioCAS). IEEE. 2012

[22] Vilathgamuwa DM, Sampath JPK. Wireless power transfer (WPT) for electric vehicles (EVs)—present and future trends. In: Plug In Electric Vehicles in Smart Grids. Singapore: Springer; 2015;**2015**:33-60

[23] Wang J et al. Optimization design of wireless charging system for autonomous robots based on magnetic resonance coupling. Journal of AIP Advances. 2018;**8**(5):055004

[24] Imura T, Hori Y. Maximizing air gap and efficiency of magnetic resonant coupling for wireless power transfer using equivalent circuit and Neumann formula. IEEE Transactions on Industrial Electronics. 2011;**58**(10):4746-4752

[25] Theodoropoulos T, Damousis Y, Amditis A. A load balancing control algorithm for EV static and dynamic wireless charging. In: IEEE 81st Vehicular Technology Conference (VTC Spring). IEEE. 2015

[26] Zhao L, Thrimawithana DJ, Madawala UK. Hybrid bidirectional wireless EV charging system tolerant to pad misalignment. IEEE Transactions on Industrial Electronics. 2017;**64**(9): 7079-7086

[27] Alexander CK, Sadiku MN, Sadiku M. Fundamentals of Electric Circuits. Boston: McGraw-Hill Higher Education; 2007

[28] Lee S, et al. On-line electric vehicle using inductive power transfer system. In: IEEE Energy Conversion Congress and Exposition. IEEE. 2010

[29] Wang K-C, et al. Study of applying contactless power transmission system to battery charge. In: 2009 International Conference on Power Electronics and Drive Systems (PEDS). IEEE. 2009

[30] Covic GA, Boys JT. Modern trends in inductive power transfer for transportation applications. IEEE Journal of Emerging Selected Topics in Power Electronics. 2013;**1**(1):28-41

[31] Lee CK, Zhong WX, Hui S. Effects of magnetic coupling of nonadjacent resonators on wireless power domino-resonator systems. IEEE Transactions on Power Electronics. 2011;**27**(4):1905-1916

[32] Lu X et al. Wireless charging technologies: Fundamentals, standards, and network applications. IEEE

Communications Surveys Tutorials. 2015;**18**(2):1413-1452

[33] Kline M. et al. Capacitive power transfer for contactless charging. In: Twenty-Sixth Annual IEEE Applied Power Electronics Conference and Exposition (APEC). IEEE. 2011

[34] Adewuyi Vo, Miantezila JM, Owoola E. Design analysis of array of dipole transmitters for wireless power transfer. Emitter International Journal of Engineering Technology. 26 April 2022: 83-101

[35] Murakami J et al. Consideration on cordless power station-contactless power transmission system. IEEE Transactions on Magnetics. 1996;**32**(5):5037-5039

[36] Hatanaka K et al. Power transmission of a desk with a cord-free power supply. IEEE Transactions on Magnetics. 2002;**38**(5):3329-3331

[37] Jayalath S, Khan A. Design, challenges, and trends of inductive power transfer couplers for electric vehicles: A review. IEEE Journal of Emerging and Selected Topics in Power Electronics. 2020;**9**(5):6196-6218

[38] Schormans MJ. Inductive Links for Biomedical Wireless Power and Data Telemetry: Circuits and Methods. UCL (University College London); 2019

[39] Schormans M, Valente V, Demosthenous A. Practical inductive link design for biomedical wireless power transfer: A tutorial. IEEE Transactions on Biomedical Circuits Systems. 2018;**12**(5):1112-1130

[40] Shi ZH, Qiu ZC, Chen XY, Li MY. Modeling and experimental verification of bidirectional wireless power transfer. 17 January 2019;**29**(2):1-5

[41] Outeiro MT, Buja G, Czarkowski D. Resonant power converters: An overview with multiple elements in the resonant tank network. IEEE Industrial Electronics Magazine. 2016;**10**(2):21-45

[42] Zhang W, Mi CC. Compensation topologies of high-power wireless power transfer systems. IEEE Transactions on Vehicular Technology. 2015;**65**(6): 4768-4778

[43] Aditya K, Williamson SS. Design considerations for loosely coupled inductive power transfer (IPT) system for electric vehicle battery charging-A comprehensive review. In: 2014 IEEE Transportation Electrification Conference and Expo (ITEC). IEEE. 2014

[44] Keeling NA, Covic GA, Boys JT. A unity-power-factor IPT pickup for high-power applications. IEEE Transactions on Industrial Electronics. 2009;**57**(2): 744-751

[45] Khaligh A, Dusmez S. Comprehensive topological analysis of conductive and inductive charging solutions for plug-in electric vehicles. IEEE Transactions on Vehicular Technology. 2012;**61**(8):3475-3489

[46] Sallán J et al. Optimal design of ICPT systems applied to electric vehicle battery charge. IEEE Transactions on Industrial Electronics. 2009;**56**(6): 2140-2149

[47] Zhang W et al. An optimized track length in roadway inductive power transfer systems. IEEE Journal of Emerging Selected Topics in Power Electronics. 2014;**2**(3):598-608

[48] Badr BM et al. Controlling wireless power transfer by tuning and detuning resonance of telemetric devices for

rodents. Wireless Power Transfer. 2020; 7(1):19-32

[49] Konishi, A., et al. Resonant frequency tuning system for repeater resonator of resonant inductive coupling wireless power transfer. In: 21st European Conference on Power Electronics and Applications (EPE'19 ECCE Europe). IEEE 2019

[50] Masuda S, et al. Impedance matching in magnetic-coupling-resonance wireless power transfer for small implantable devices. In: IEEE Wireless Power Transfer Conference (WPTC). IEEE. 2017

[51] Beh TC, Imura T, Kato M, Hori Y. Basic study of improving efficiency of wireless power transfer via magnetic resonance coupling based on impedance matching. IEEE International Symposium on Industrial Electronics. Jul 4 2010:2011-2016

[52] Sample AP, Meyer DT, Smith JR. Analysis, experimental results, and range adaptation of magnetically coupled resonators for wireless power transfer. IEEE Transactions on Industrial Electronics. 2010;**58**(2): 544-554

[53] Duong TP, Lee J-W. Experimental results of high-efficiency resonant coupling wireless power transfer using a variable coupling method. IEEE Microwave Wireless Components Letters. 2011;**21**(8):442-444

[54] Hui SYR, Zhong W, Lee CK. A critical review of recent progress in mid-range wireless power transfer. IEEE Transactions on Power Electronics. 2013; **29**(9):4500-4511

[55] Kiani M, Jow U-M, Ghovanloo M. Design and optimization of a 3-coil inductive link for efficient wireless power transmission. IEEE Transactions on Biomedical Circuits Systems. 2011; **5**(6):579-591

[56] Shevchenko V et al. Compensation topologies in IPT systems: Standards, requirements, classification, analysis, comparison and application. IEEE Access. 2019;7:120559-120580

[57] Huang Z, Wong S-C, Chi KT. Comparison of basic inductive power transfer systems with linear control achieving optimized efficiency. IEEE Transactions on Power Electronics. 2019; **35**(3):3276-3286

[58] Movagharnejad H, Mertens A. Design metrics of compensation methods for contactless charging of electric vehicles. In: 19th European Conference on Power Electronics and Applications (EPE'17 ECCE Europe). IEEE. 2017

[59] Bosshard R, Kolar JW, M¨hlethaler J, Stevanović I, Wunsch B, Canales F. Modeling and ε-α-Pareto Optimization of Inductive Power Transfer Coils for Electric Vehicles. IEEE Journal of Emerging and Selected Topics in Power Electronics. 2014 Mar 11;**3**(1):50-64

[60] Choi SY et al. Generalized active EMF cancel methods for wireless electric vehicles. IEEE Transactions on Power Electronics. 2013;**29**(11): 5770-5783

[61] Kim S et al. Design and analysis of a resonant reactive shield for a wireless power electric vehicle. IEEE Transactions on Microwave Theory Techniques. 2014;**62**(4):1057-1066

[62] Wang C-S, Stielau OH, Covic GA. Design considerations for a contactless electric vehicle battery charger. IEEE Transactions on Industrial Electronics. 2005;**52**(5):1308-1314

[63] Villa JL et al. High-misalignment tolerant compensation topology for ICPT systems. IEEE Transactions on Industrial Electronics. 2011;**59**(2):945-951

[64] Samanta S, Rathore AK. Wireless power transfer technology using full-bridge current-fed topology for medium power applications. IET Power Electronics. 2016;**9**(9):1903-1913

[65] Tsai J-S et al. Directional antenna design for wireless power transfer system in electric scooters. Advances in Mechanical Engineering. 2016;**8**(2): 1687814016632693

[66] Fu M, Tang Z, Ma C. Analysis and optimized design of compensation capacitors for a megahertz WPT system using full-bridge rectifier. IEEE Transactions on Industrial Informatics. 2018;**15**(1):95-104

[67] Covic GA, Boys JT. Inductive power transfer. Proceedings of the IEEE. 2013; **101**(6):1276-1289

[68] Cai C et al. Design and optimization of load-independent magnetic resonant wireless charging system for electric vehicles. IEEE Access. 2018;**6**: 17264-17274

[69] Li S et al. A double-sided LCC compensation network and its tuning method for wireless power transfer. IEEE Transactions on Vehicular Technology. 2014;**64**(6):2261-2273

[70] Yao Y et al. Analysis, design, and optimization OFLC/S compensation topology with excellent load-independent voltage output for inductive power transfer. IEEE Transactions on Transportation Electrification. 2018;**4**:767-777

[71] Chigira M, et al. Small-size light-weight transformer with new core structure for contactless electric vehicle power transfer system. In: IEEE Energy Conversion Congress and Exposition. IEEE. 2011

[72] Aworo OJ, Shek JK. Transformer for contactless electric vehicle charging with bidirectional power flow. In: IEEE Power & Energy Society General Meeting. IEEE. 2017

[73] Kan T et al. A new integration method for an electric vehicle wireless charging system using LCC compensation topology: Analysis and design. IEEE Transactions on Industrial Electronics. 2016;**32**(2):1638-1650

[74] Rasekh N, Kavianpour J, Mirsalim M. A novel integration method for a bipolar receiver pad using LCC compensation topology for wireless power transfer. IEEE Transactions on Vehicular Technology. 2018;**67**(8):7419-7428

[75] Li Y et al. Compact double-sided decoupled coils-based WPT systems for high-power applications: Analysis, design and experimental verification. IEEE Transactions on Transportation Electrification;**4**(1):64-75

[76] Li W et al. Comparison study on SS and double-sided LCC compensation topologies for EV/PHEV wireless chargers. IEEE Transactions on Vehicular Technology. 2015;**65**(6):4429-4439

[77] Lu F et al. A dynamic charging system with reduced output power pulsation for electric vehicles. IEEE Transactions on Industrial Electronics. 2016;**63**(10):6580-6590

[78] Li W et al. Integrated LCC compensation topology for wireless charger in electric and plug-in electric vehicles. IEEE Transactions on Industrial Electronics. 2014;**62**(7):4215-4225

[79] Mou X et al. Survey on magnetic resonant coupling wireless power transfer technology for electric vehicle charging. IET Power Electronics. 2019; **12**(12):3005-3020

[80] Ku M-L et al. Power waveforming: Wireless power transfer beyond time reversal. IEEE Transactions on Signal Processing. 2016;**64**(22):5819-5834

[81] Hu Z, Yuan C, Gao F. Maximizing harvested energy for full-duplex SWIPT system with power splitting. IEEE Access. 2017;**5**:24975-24987

[82] Huang Y, Clerckx B. Large-scale multiantenna multisine wireless power transfer. IEEE Transactions on Signal Processing. 2017;**65**(21):5812-5827

[83] Sun H et al. Magnetic resonant beamforming for secured wireless power transfer. IEEE Signal Processing Letters. 2017;**24**(8):1173-1177

[84] Lang H-D, Ludwig A, Sarris CD. Optimization and design sensitivity of SISO and MISO wireless power transfer systems. In: IEEE International Symposium on Antennas and Propagation & USNC/URSI National Radio Science Meeting. IEEE. 2015

[85] Agcal A, Ozcira S, Bekiroglu N. Wireless power transfer by using magnetically coupled resonators. Journal of Wireless Power Transfer: Fundamentals Technologies. 2016:49-66

[86] Junussov A, Bagheri M, Lu M. Analysis of magnetically coupled resonator and four-coil wireless charging systems for EV. In: International Conference on Sustainable Energy Engineering and Application (ICSEEA). IEEE. 2017

[87] Huang S, Li Z, Lu K. Frequency splitting suppression method for four-coil wireless power transfer system. IET Power Electronics. 2016;**9**(15):2859-2864

[88] Sultanbek A, et al. Intelligent wireless charging station for electric vehicles. In: International Siberian Conference on Control and Communications (SIBCON). IEEE. 2017

[89] Moon S, Moon G-W. Wireless power transfer system with an asymmetric four-coil resonator for electric vehicle battery chargers. IEEE Transactions on Power Electronics. 2015;**31**(10): 6844-6854

[90] Kang SH, Choi JH, Jung CW. Magnetic resonance wireless power transfer using three-coil system with single planar receiver for laptop applications. IEEE Transactions on Consumer Electronics. 2015;**61**(2): 160-166

[91] Jegadeesan R, Guo Y-X. Modeling of wireless power transfer link for retinal implant. In: IEEE/ACES International Conference on Wireless Information Technology and Systems (ICWITS) and Applied Computational Electromagnetics (ACES). IEEE. 2016

[92] Jegadeesan R et al. Enabling wireless powering and telemetry for peripheral nerve implants. IEEE Journal of Biomedical Health Informatics. 2015; **19**(3):958-970

[93] Kesler M. Wireless charging of electric vehicles. In: IEEE Wireless Power Transfer Conference (WPTC). 2018

[94] Beeton D, Holland B. EV city casebook. In: 50 Big Ideas Shaping the Future of Electric Mobility. EVI, IA-HEV: Urban Foresight Limited; 2014

[95] Shin J et al. Design and implementation of shaped magnetic-

resonance-based wireless power transfer system for roadway-powered moving electric vehicles. IEEE Transactions on Industrial Electronics. 2013;**61**(3): 1179-1192

[96] Eghtesadi M. Inductive power transfer to an electric vehicle-analytical model. In: 40th IEEE Conference on Vehicular Technology. IEEE. 1990

[97] Kim H et al. Coil design and measurements of automotive magnetic resonant wireless charging system for high-efficiency and low magnetic field leakage. IEEE Transactions on Microwave Theory Techniques. 2016; **64**(2):383-400

[98] Kim J, et al. Efficiency of magnetic resonance WPT with two off-axis self-resonators. In: IEEE MTT-S International Microwave Workshop Series on Innovative Wireless Power Transmission: Technologies, Systems, and Applications. IEEE. 2011

[99] Zhang W et al. Loosely coupled transformer structure and interoperability study for EV wireless charging systems. IEEE Transactions on Power Electronics. 2015;**30**(11): 6356-6367

[100] Xianjin S, et al. Analyses and experiments of field-circuit coupling equations for wireless power transfer using solenoidal coils. In: IEEE International Wireless Symposium (IWS 2015). IEEE. 2015

[101] Feenaghty M, Dahle R. A compact and high quality factor archimedean coil geometry for wireless power transfer. In: IEEE Wireless Power Transfer Conference (WPTC). IEEE. 2016

[102] Flynn BW, Fotopoulou K. Rectifying loose coils: Wireless power transfer in loosely coupled inductive links with lateral and angular misalignment. IEEE Microwave Magazine. 2013;**14**(2):48-54

[103] Pantic Z, Lukic S. Computationally-efficient, generalized expressions for the proximity-effect in multi-layer, multi-turn tubular coils for wireless power transfer systems. IEEE Transactions on Magnetics. 2013;**49**(11):5404-5416

[104] Budhia M et al. Development of a single-sided flux magnetic coupler for electric vehicle IPT charging systems. IEEE Transactions on Industrial Electronics. 2011;**60**(1):318-328

[105] Zaheer A, et al. Magnetic design of a 300 W under-floor contactless power transfer system. In: 37th Annual Conference of the IEEE Industrial Electronics Society. IEEE. 2011

[106] Kim S, Covic GA, Boys JT. Comparison of tripolar and circular pads for IPT charging systems. IEEE Transactions on Power Electronics. 2017; **33**(7):6093-6103

[107] Benomar Y, et al. Design and modeling of V2G inductive charging system for light-duty Electric Vehicles. In: Twelth International Conference on Ecological Vehicles and Renewable Energies (EVER). IEEE. 2017

[108] Kim S, Covic GA, Boys JT. Tripolar pad for inductive power transfer systems for EV charging. IEEE Transactions on Power Electronics. 2016;**32**(7):5045-5057

[109] Tejeda A et al. A hybrid solenoid coupler for wireless charging applications. IEEE Transactions on Power Electronics. 2018;**34**(6):5632-5645

[110] Mecke R, Rathge C. High frequency resonant inverter for contactless energy transmission over large air gap. In: IEEE 35th Annual Power Electronics

Specialists Conference (IEEE Cat. No. 04CH37551). IEEE 2004

[111] Li Y. et al. Wireless energy transfer system based on high Q flexible planar-Litz MEMS coils. In: The 8th Annual IEEE International Conference on Nano/ Micro Engineered and Molecular Systems. IEEE. 2013

[112] Mizuno T et al. Improvement in efficiency of wireless power transfer of magnetic resonant coupling using magnetoplated wire. IEEE Transactions on Magnetics. 2011;**47**(10):4445-4448

[113] Sampath JPK, et al. Analysis on normalized distance and scalability in designing wireless power transfer. In: Proc. IEEE PELS Workshop on Emerging Technologies: Wireless Power (WoW). 2015. pp. 1-6

[114] Kurschner D, Rathge C, Jumar U. Design methodology for high efficient inductive power transfer systems with high coil positioning flexibility. IEEE Transactions on Industrial Electronics. 2011;**60**(1):372-381

[115] Miller JM, Onar OC, Chinthavali M. Primary-side power flow control of wireless power transfer for electric vehicle charging. IEEE Journal of Emerging Selected Topics in Power Electronics. 2014;**3**(1):147-162

[116] Bosshard R. Multi-objective optimization of inductive power transfer systems for EV charging. PhD diss, ETH Zurich, 2015

[117] Bosshard R et al. Modeling and ε-α-Pareto optimization of inductive power transfer coils for electric vehicles. IEEE Journal of Emerging. 2014;**3**(1):50-64

[118] Miller JM, Daga A. Elements of wireless power transfer essential to high power charging of heavy duty vehicles. IEEE Transactions on Transportation Electrification. 2015;**1**(1):26-39

[119] Bosshard R, Kolar JW. Multi-objective optimization of 50 kW/ 85 kHz IPT system for public transport. IEEE Journal of Emerging Selected Topics in Power Electronics. 2016;**4**(4): 1370-1382

[120] Budhia M, Covic GA, Boys JT. Design and optimization of circular magnetic structures for lumped inductive power transfer systems. IEEE Transactions on Power Electronics. 2011; **26**(11):3096-3108

[121] Zaheer A, Covic GA, Kacprzak D. A bipolar pad in a 10-kHz 300-W distributed IPT system for AGV applications. IEEE Transactions on Industrial Electronics. 2013;**61**(7): 3288-3301

[122] Ning P, Wen X. Genetic algorithm based coil system design for wireless power charging. In: IEEE Applied Power Electronics Conference and Exposition-APEC 2014. IEEE. 2014

[123] Hasan N, et al. Multi-objective particle swarm optimization applied to the design of wireless power transfer systems. In: IEEE Wireless Power Transfer Conference (WPTC). IEEE. 2015

[124] Liu Y et al. Efficiency optimization for wireless dynamic charging system with overlapped DD coil arrays. IEEE Transactions on Power Electronics. 2017; **33**(4):2832-2846

[125] Sampath JPK, Alphones A, Vilathgamuwa DM. Figure of merit for the optimization of wireless power transfer system against misalignment tolerance. IEEE Transactions on Power Electronics. 2016;**32**(6):4359-4369

[126] Desmoort A, De Grève Z, Deblecker O. Multiobjective optimal design of wireless power transfer devices using a genetic algorithm and accurate analytical formulae. In: IECON 2016-42nd Annual Conference of the IEEE Industrial Electronics Society. IEEE. 2016

[127] Castillo-Zamora IU et al. Hexagonal geometry coil for a WPT high-power fast charging application. IEEE Transactions on Transportation Electrification. 2019; **5**(4):946-956

[128] Hariri A, Elsayed A, Mohammed OA. An integrated characterization model and multiobjective optimization for the design of an EV charger's circular wireless power transfer pads. IEEE Transactions on Magnetics. 2017;**53**(6): 1-4

[129] Blanco S. Sae now has a wireless charging standard: j2954. June 6, 2022; Available from: https://www.sae.org/news/2020/10

Chapter 5

Iterative Technique for Analysis and Design of Circular Leaky-Wave Antenna for the 2.45 GHz RFID Applications

Nizar Sghaier

Abstract

This chapter proposes another scaled-down receiving wire for Radio Frequency Identification uses of 2.45 GHz. Our design comprises of a roundabout microstrip fix radio wire which integrates two concentric annular openings imprinted on multi-facet substrates. The transmission capacity, one of the main qualities of radio wire, can be essentially improved by utilizing a multi-facet dielectric setup. We point by this review to show that the impact of emanating structure stacked by annular rings for the fixed size decrease as well as the benefit of this construction is to make a roundabout polarization toward maximal radiation design. The wave idea iterative method is utilized to examine this new radio wire. Utilizing the proposed technique, less figuring time and memory are expected to work out the electromagnetic boundaries of our plan. The approval of the consequences of our created model was checked with realized business programming called "CST Microwave Studio Software" trailed by a trial test. As per the arrived results, we can decide that our new plan radio wire is reasonable for RFID applications in the 2.45 GHz band.

Keywords: RFID, concentric annular slots, multilayer substrates, bandwidth, WCIP, radiation pattern

1. Introduction

The exponential evolution of the need for microwave devices in modern communication systems has prompted manufacturers to invest more and more in this field in order to meet recent market trends. These systems are often multi-band in order to meet several communication standards. Their multiplicity on the same carrier means that more and more attention is being paid to reducing their size. The antenna is one of the essential components in wireless systems for both civil and military applications, space and terrestrial [1–4].

The diversity of the fields of activity using these antennas has increased the need to develop antennas that are agile (in frequency, pattern, or polarization) and multistandard while keeping a compact appearance. The radiation of a circularly polarized wave is often of interest in order to overcome depolarization phenomena that can occur during propagation [5–8].

In order to meet the various challenges, a multitude of avenues has been explored to adapt to the performance requirements of the applications. Indeed, various topologies and techniques have been investigated in order to meet the required specifications, namely: miniaturization, multi-band operation, and circular polarization radiation while maintaining optimal radio performances (gain, radiation efficiency, reflection coefficient). In addition, a promising way to improve the performance of an antenna is to integrate innovative materials. Many research works have explored this way. Our objective is, therefore, to design miniature microstrip antenna structures for RFID readers in the UHF band with the correct performance [9–11].

Circularly polarized antennas are the most commonly used antennas in RFID communications. In addition, for optimal operation, these antennas must have a particular radiation pattern. It must be omnidirectional in any azimuthal cutting plane, and maintain a high gain, although the choice of a circular antenna is justified by the fact that it has the advantage of being able to achieve circular polarization [12–20].

The rigorous analysis and design of microwave circuits require first of all a rigorous resolution of the equations governing the electromagnetic field, in order to be able to reconstruct the closest to real behavior of the fields in the devices. To satisfy this need, in this chapter, we focus on the study of the circular antenna by the wave concept iterative method (W.C.I.P). This method is based on the manipulation of the incident and reflected waves instead of the electromagnetic field. It is based on the back and forth between the spatial and spectral domains, using the Fourier transform FMT followed by the Hankel transform in the cylindrical coordinate system [21–26].

2. WCIP presentation

2.1 Formulation

To begin, **Figure 1** shows the starting structure.

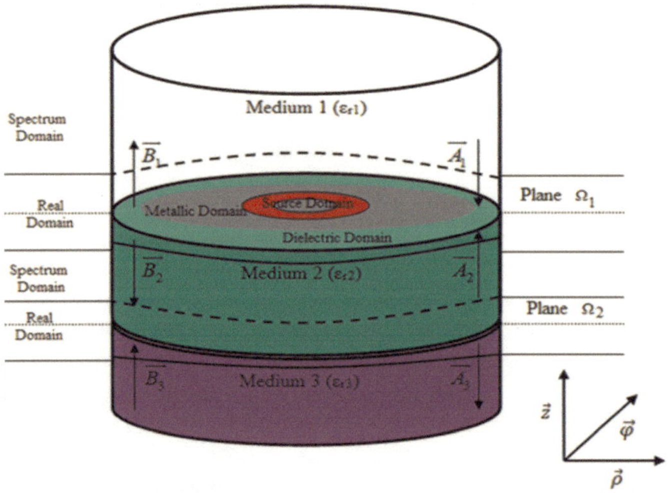

Figure 1.
WCIP structure.

The incident wave and the scattered waves are calculated from the tangential magnetic and electric field at the surface:

$$\begin{bmatrix} A_i \\ B_i \end{bmatrix} = \frac{1}{2\sqrt{Z_{0i}}} \begin{bmatrix} 1 & \sqrt{Z_{0i}} \\ 1 & -\sqrt{Z_{0i}} \end{bmatrix} \begin{bmatrix} E_i \\ J_i \end{bmatrix} \tag{1}$$

Where B_i and A_i are the incident and reflected waves associated with the discontinuity plane Ω_1. Z_{0i} is the characteristic impedance of the medium i (i = 1, 2) given by $Z_{0i} = \sqrt{\mu_0/\varepsilon_0\varepsilon_{ri}}$ (**Figure 2**).

The steps of the iterative method are:

1. Define ($\overrightarrow{B_0}$).
2. Calculates $\overrightarrow{B_{\rho,\varphi}} = \Gamma_{\Omega_1}\overrightarrow{A_{\rho,\varphi}} + \overrightarrow{B_0}$
3. Apply HT and FFT: $HT_{/\rho}\left(FFT_{/\varphi}\left(\overrightarrow{B_{\rho,\varphi}}\right)\right) = \overrightarrow{B^{TE,TM}_{m,n}}$
4. Calculates $\overrightarrow{A^{TE,TM}_{m,n}} = \Gamma_i \overrightarrow{B^{TE,TM}_{m,n}}$
5. Apply IHT and IFFT: $HT_{/\rho\text{-}1}\left(FFT_{/\varphi\text{-}1}\left(\overrightarrow{A^{TE,TM}_{m,n}}\right)\right) = \left(\overrightarrow{A_{\rho,\varphi}}\right)$
6. Repeat steps 2 to 5.

$$\begin{cases} E^k_{/\varphi,\rho} = \sqrt{Z_{0i}}\left(A^k_i + B^k_i\right) \\ J^k_{/\varphi,\rho} = \left(A^k_i + B^k_i\right)/\sqrt{Z_{0i}} \end{cases} \tag{2}$$

The expression of return loss at the upper and bottom side of box in the spectrum domain is given by:

$$\hat{\Gamma}_i^{TE,TM} = \frac{1 - Z_{0i}\hat{Y}^{TE,TM}_{m,n}}{1 + Z_{0i}\hat{Y}^{TE,TM}_{m,n}} \tag{3}$$

Where the admittance $\hat{Y}^{TE,TM}_{m,n}$ is done by **Table 1**:

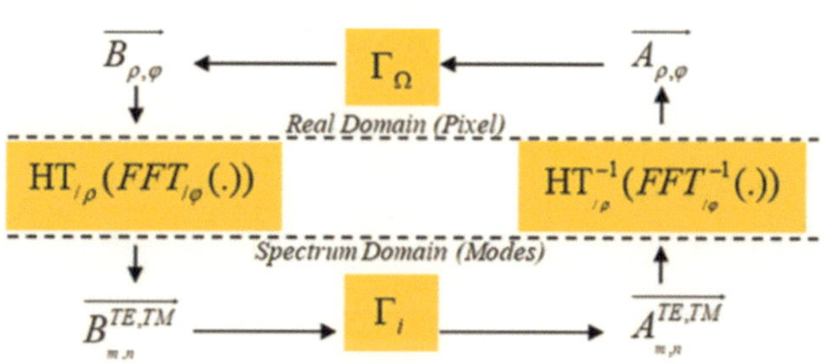

Figure 2.
WCIP procedure.

	Mode TE	Mode TM
$\hat{Y}^{TE,TM}_{m,n}$	$-jY_r \frac{\gamma}{K_r} \coth(\gamma h)$	$jY_r \frac{K_r}{\gamma} \coth(\gamma h)$

Table 1.
Admittance expressions.

$k_r = \sqrt{\omega^2 \varepsilon_0 \varepsilon_r \mu_0}$, $\gamma = k_\rho^2 - k_r^2$ *and* $Y_r = \sqrt{\frac{\varepsilon_0 \varepsilon_r}{\mu_0}}$ are the admittance of each domain.

The equivalent circuit of the model is shown in the figure below:
see (**Figure 3**).

The coupling between two layers is characterized by the equivalent admittance Y. The modal admittance seen at the interface Ω_i between layers i-2 and i-1 can be calculated by:

$$Y^{TE,TM}_{m,n}(\mathrm{i}-1) = \hat{Y}^{TE,TM}_{m,n}(\mathrm{i}-1)\left(\frac{Y^{TE,TM}_{m,n}(\mathrm{i}-2) + \hat{Y}^{TE,TM}_{m,n}(\mathrm{i}-1)\tanh\left(\gamma^{i-1}_{m,n} h_{i-1}\right)}{\hat{Y}^{TE,TM}_{m,n}(\mathrm{i}-1) + Y^{TE,TM}_{m,n}(\mathrm{i}-2)\tanh\left(\gamma^{i-1}_{m,n} h_{i-1}\right)}\right), \quad \mathrm{i} = 2{,}3 \tag{4}$$

At the printed surface of the discontinuity, the boundary conditions of fields are expressed in terms of waves that consist of three conditions as:

The boundary conditions are given by (5):

$$\begin{cases} E_1 = E_2 = 0 & (Metal) \\ J_1 + J_2 = 0 & (Dielectric) \\ E = E_0 - z_0(J_1 + J_2) & (Source) \end{cases} \tag{5}$$

The relationship between the incident and reflected waves in the spatial domain is given by:

$$B_i = \hat{\Gamma}_\Omega A_i + B_0 \tag{6}$$

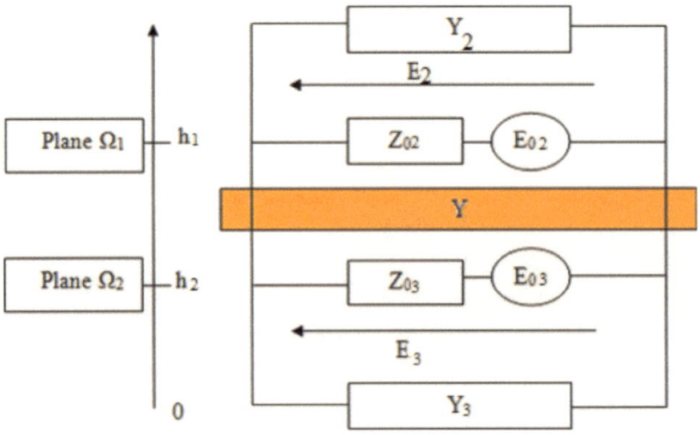

Figure 3.
Equivalent circuit of multilayered structure.

With B_0 is the source excitation, the diffraction operator Γ_Ω is defined by:

$$\Gamma_\Omega = \begin{pmatrix} \Gamma_{\Omega 11} & \Gamma_{\Omega 12} \\ \Gamma_{\Omega 21} & \Gamma_{\Omega 22} \end{pmatrix} \tag{7}$$

Where:

$$\begin{cases} \Gamma_{\Omega 11} = -H_m - \dfrac{-1+n_1+n_2}{1+n_1+n_2} H_s + \dfrac{1-n^2}{1+n^2} H_d \\ \Gamma_{\Omega 12} = \dfrac{2n}{1+n^2} H_d + \dfrac{2n}{1+n_1+n_2} H_s \\ \Gamma_{\Omega 21} = \dfrac{2n}{1+n^2} H_d + \dfrac{2n}{1+n_1+n_2} H_s \\ \Gamma_{\Omega 22} = -H_m - \dfrac{-1-n_1+n_2}{1+n_1+n_2} H_s + \dfrac{1-n^2}{1+n^2} H_d \end{cases}$$

Where H is Heaviside function (m: metal; d: dielectric; S: source)

$$\text{And n} = \frac{Z0}{\sqrt{Z_{01} Z_{02}}}, n_1 = \frac{Z_0}{Z_{01}}, n_2 = \frac{Z_0}{Z_{02}}; \tag{8}$$

The passage between the spectral and spatial domain is given by the following **Table 2**:

The radial cut-off constants and normalisation constants for the modes and are given in **Table 3**:

The passage into the spectral domain then requires the transformation of Hankel's incident waves [27]:

The passage into the spectral domain then requires the Hankel transform of the incident waves [7]:

Mode TE	Mode TM
$E_\rho = \frac{e^{j\varphi}}{\sqrt{\pi}} \frac{1}{\Lambda\rho} J_1(\kappa_{\rho m})$	$E_\rho = -j \frac{e^{j\varphi}}{\sqrt{\pi}} \frac{\kappa_{\rho m}}{\Lambda'} J_1'(\kappa_{\rho m}\rho)$
$E_\varphi = j \frac{e^{j\varphi}}{\sqrt{\pi}} \frac{\kappa_{\rho m}}{\Lambda} J_1'(\kappa_{\rho m}\rho)$	$E_\varphi = \frac{e^{j\varphi}}{\sqrt{\pi}} \frac{1}{\Lambda'\rho} J_1(\kappa_{\rho m})$

Table 2.
Standards mode.

	TE and electric wall Or TM and magnetic wall	TM and electric wall Or TE and magnetic wall
$\kappa_{\rho m}$	$J_1(\kappa_\rho a) = 0$	$J_1'(\kappa_\rho a) = 0$
Λ, Λ'	$\Lambda^2 = \frac{1}{2}\kappa_\rho^2 a^2 J_1'^2(\kappa_{\rho m} a)$	$\Lambda'^2 = \frac{1}{2}\left(\kappa_{\rho m}^2 a^2 - 1\right) J_1^2(\kappa_{\rho m} a)$

Table 3.
Cut-off and modes normalisation.

$$\begin{pmatrix} B^{TE} \\ B^{TM} \end{pmatrix} = \frac{e^{j\varphi}}{\sqrt{2\pi}} \begin{pmatrix} \frac{1}{\Lambda\rho} TH_1 & j\frac{\kappa_\rho}{\Lambda} TH_1' \\ -j\frac{\kappa_\rho'}{\Lambda'} TH_1' & \frac{1}{\Lambda'\rho} TH_1 \end{pmatrix} \begin{pmatrix} B_\rho \\ B_\varphi \end{pmatrix} \tag{9}$$

With κ and κ′ are the zeros of the Bessel functions and its first-order derivative, and are the normalisation constants of the TE and TM modes respectively and are given by:

$$\begin{cases} TH_1(f(\rho)) = \int_0^a f(\rho) J_1(k_\rho \rho) \rho d\rho \\ TH_{1'}(f(\rho)) = \int_0^a f(\rho) J_{1'}(k_\rho \rho) \rho d\rho \end{cases} \tag{10}$$

The transition from the spectral domain to the space domain is mandatory. This passage is done by the inverse Hankel transform [7]. We note and respectively follows:

$$\begin{cases} TH_{1\_1}(f(\kappa_\rho)) = \int_0^a f(\kappa_\rho) J_1(k_\rho \rho) \kappa_\rho d\kappa_\rho \\ TH'_{1\_1}(f(\kappa_\rho)) = \int_0^a f(\kappa_\rho) J_{1'}(k_\rho \rho) \kappa_\rho d\kappa_\rho \end{cases} \tag{11}$$

The transition to the spatial domain takes the following matrix form:

$$\begin{pmatrix} B_\rho \\ B_\varphi \end{pmatrix} = \frac{e^{j\varphi}}{\sqrt{2\pi}} \begin{pmatrix} TH_1^{-1}\left(\frac{1}{\Lambda\rho}\right. & TH_1'^{-1}\left(j\frac{\kappa_\rho}{\Lambda}\right. \\ TH_1'^{-1}\left(-j\frac{\kappa_\rho'}{\Lambda'}\right. & TH_1^{-1}\left(\frac{1}{\Lambda'\rho}\right) \end{pmatrix} \begin{pmatrix} B^{TE} \\ B^{TM} \end{pmatrix} \tag{12}$$

The numerical implementation of the developed method is given by the flowchart below (**Figure 4**):

2.2 Validation

Figure 5 shows the structure of a patch antenna to validate the method which is already developed. This antenna is engraved on a dielectric with relative permittivity $\varepsilon_{r1} = 4{,}25$. The second dielectric is of relative permittivity $\varepsilon_{r2} = 3{,}3$. The dimensions of the patch are given in **Table 4**. We use a discretisation of 100*40 pixels (100 pixels for the radial direction and 40 for the ortho-radial direction).

In order to validate the iterative method, we started first with the comparison of the operation's number between WCIP and MOM which the most used in scientific research [28–30].

The operation number of WCIP is given by:

$$N_{OP} = N\left(4P + 12\log_2 P\right) \tag{13}$$

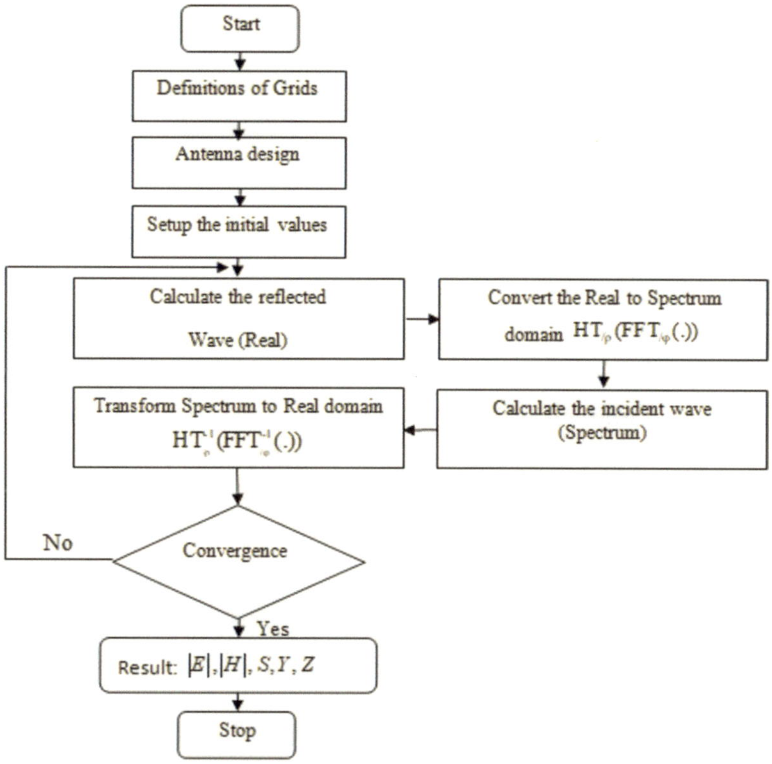

Figure 4.
Flowchart of the WCIP simulation.

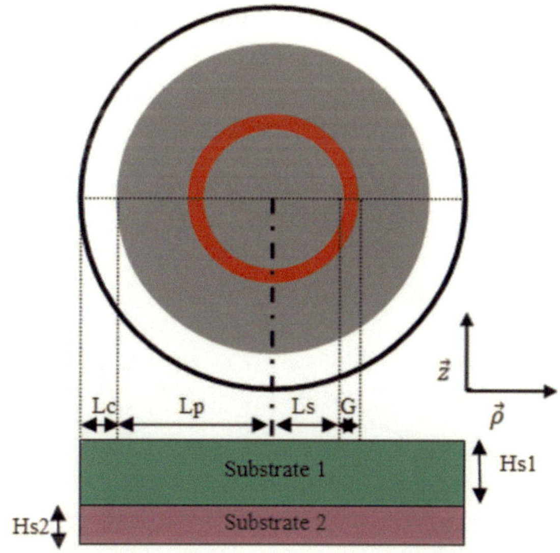

Figure 5.
Circular patch excited by a variable source.

Parameter	Value (mm)
Lc	1
Lp	17
Ls	5.4
G	0.8
Hs1	2
Hs2	1

Table 4.
Dimensions of the structure.

The operation number of MOM is given by:

$$N_{OP} = (KP)^3/3 \tag{14}$$

Where:

- P is the number of meshes
- N is the number of iterations
- K is the proportion of metallic surface in the circuit structure.

Figure 6 shows the number of operations required for WCIP and MOM (**Figure 7**).
The convergence of WCIP is completed at 22 iterations as shown in the figure above.
A comparative study is made between the developed method and the MOM method as well as between these two methods and two software CST and HFSS.
Based on **Figure 8**, it can be deduced that the iterative method is the most significant in terms of accuracy.
Table 5 shows the simulation time and memory consumption of each method.

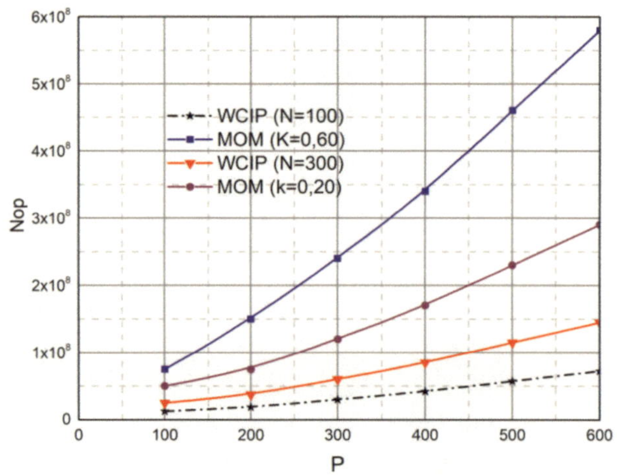

Figure 6.
Operation's number between WCIP and MOM.

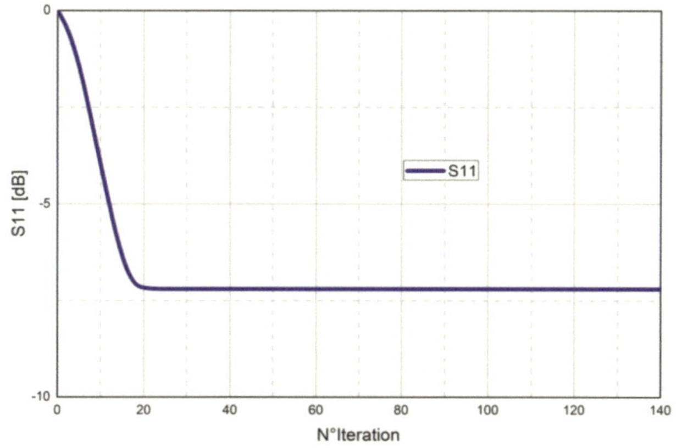

Figure 7.
Convergence versus iterations number at 2.35 GHz.

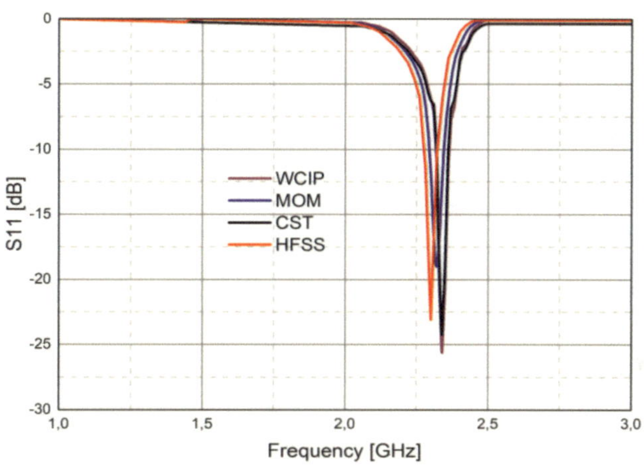

Figure 8.
Simulation return loss using WCIP, MOM, CST and HFSS.

Methods	Response time (s)	Consumption memory (MB)
WCIP	480	450
CST	2100	620
MOM	1756	905
HFSS	629	479

Table 5.
Comparison between these methods.

3. Antenna design

Figure 9 depicts the geometry of the circular leaky-wave antenna. It is characterized by a circular patch of radius Lp = 20 mm which incorporates two concentric annular slots of widths G1 = G2 = 1.058 mm and of successive inner radius R1 = 7.058 mm

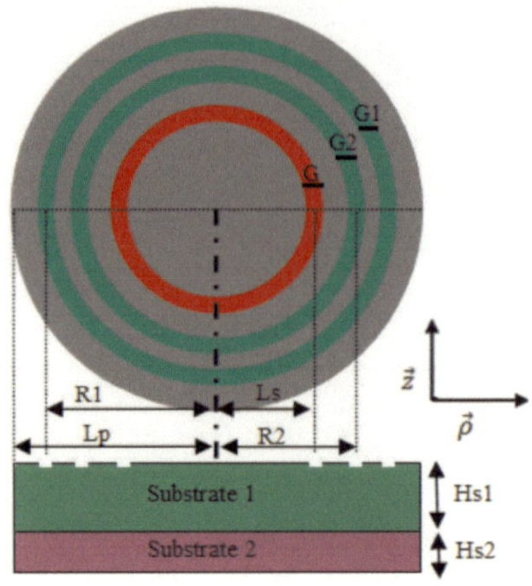

Figure 9.
Antenna design.

and R2 = 12.882 mm is printed on multilayer substrates. The first substrate layer is characterized by a height H_{s1} = 1 mm and a permittivity ε_{r1} = 4.25. The parameters (H_{s2}, ε_{r2}) of the second layer will be optimized later. The antenna is supplied by an annular planar source of inner radius Ls = 1.941 mm and of width G = 1.058 mm.

To fix the dimensions of the antenna, a parametric study is necessary. **Figures 10–13** show the variation of the antenna performance as a function of the height of the antenna substrate 2 and its permittivity.

The parametric study shows that the best compromise between gain, efficiency and bandwidth at 2.45 GHz is obtained when H_{s2} = 0.3 mm and ε_{r2} = 2.2.

4. Results and discussions

The results presented in **Figures 14** and **15** show that the antenna has a resonance peak at the frequency 2.45GHz which corresponds to -38.32 dB with a bandwidth of

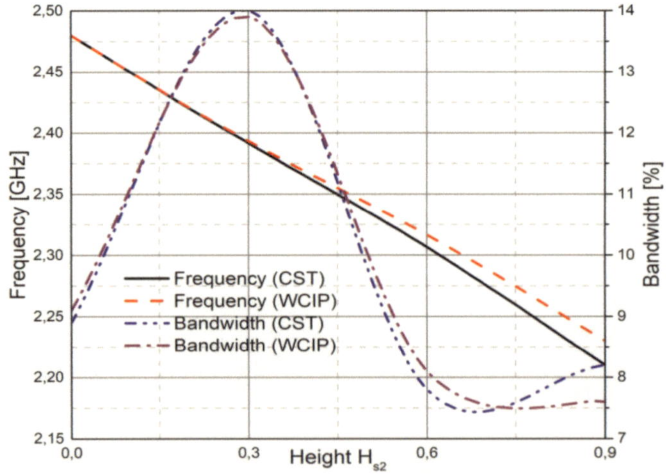

Figure 10.
Simulation frequency and bandwidth for different height of substrate 2 using WCIP and CST.

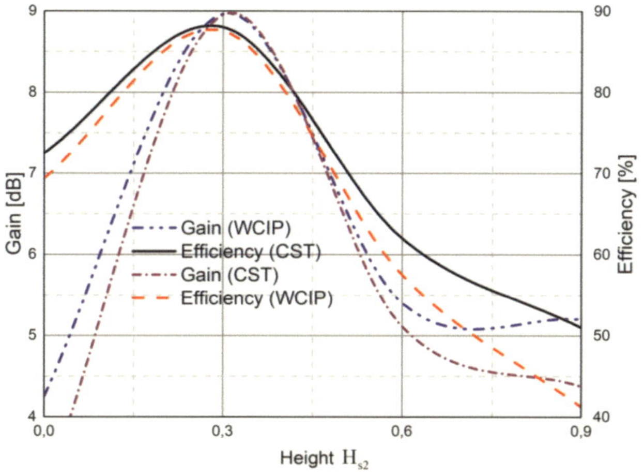

Figure 11.
Simulation gain and efficiency for different height of substrate 2 using WCIP and CST.

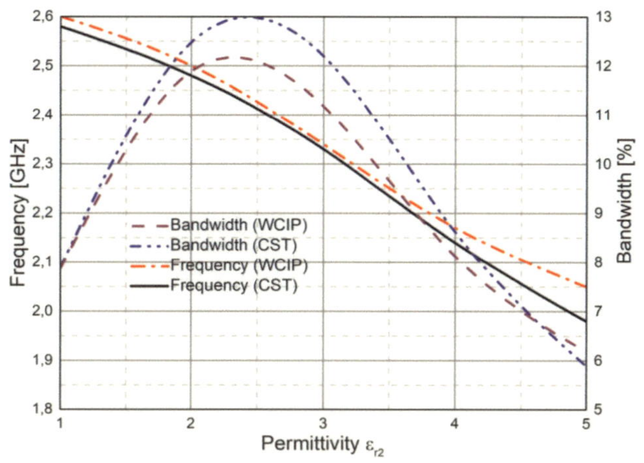

Figure 12.
Simulation frequency and bandwidth for different permittivity of substrate 2 using WCIP and CST.

310MHz (ranging from 2.29 to 2.6GHz) as well as the measured resonance peak is 2.45GHz with –28.57 dB reflection coefficient, which cover the bandwidth of 350MHz (from 2.28 to 2.63GHz). So, the antenna is applicable for RFID applications at 2.45 GHz. In this proposed design, we notice a good agreement between the simulated and measured results.

The measured and simulated radiation patterns at the resonance frequency are given by **Figure 16**. Following the analysis of the results the phi-planes show mostly unidirectional radiation patterns over the specified frequencies.

Figure 16 shows the measured and simulated gain versus the frequency of the proposed antenna. From the simulated results the maximum value of gain is 9.02dB at 2.45GHz as well as the measured value is 8.96dB at 2.45GHz.

The efficiency of the antenna is depicted in **Figure 17**. We obtain a satisfactory value of 88% over a wide frequency bandwidth.

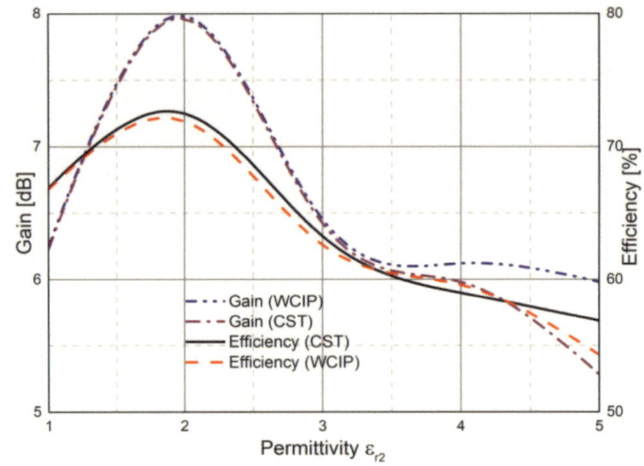

Figure 13.
Simulation gain and efficiency for different permittivity of substrate 2 using WCIP and CST.

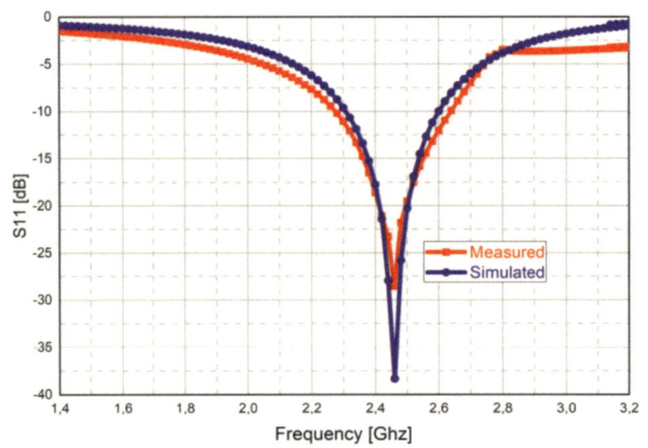

Figure 14.
Simulation and measurement return loss using WCIP.

In free-space, the power received by a tag antenna P_a can be calculated using the Friis free-space [31] equation, where:

$$P_a = \frac{P_r G_r G_a \lambda^2}{(4\pi)^2 d^2} \tag{15}$$

Where (**Table 6**):
The Eq. (16) represents the read range ®:

$$r = \frac{\lambda}{4\pi}\sqrt{\frac{P_r G_r G_a \tau}{P_{th}}} \tag{16}$$

According to the results of our structure at 2.45 GHz, the numerical values are given in the **Table 7**:

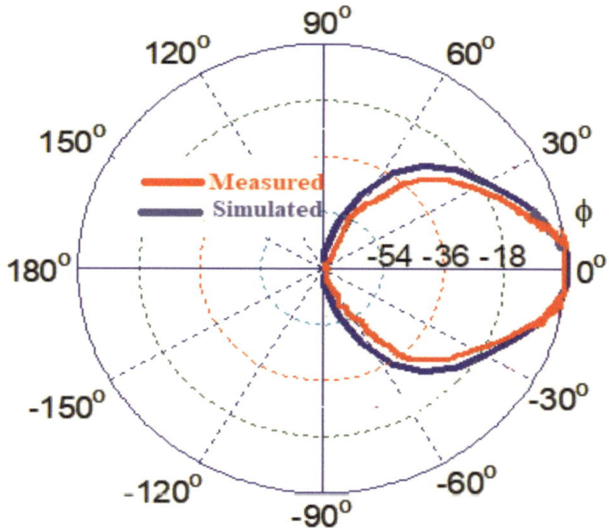

Figure 15.
Simulation and measurement radiation pattern at 2.45GHz using WCIP.

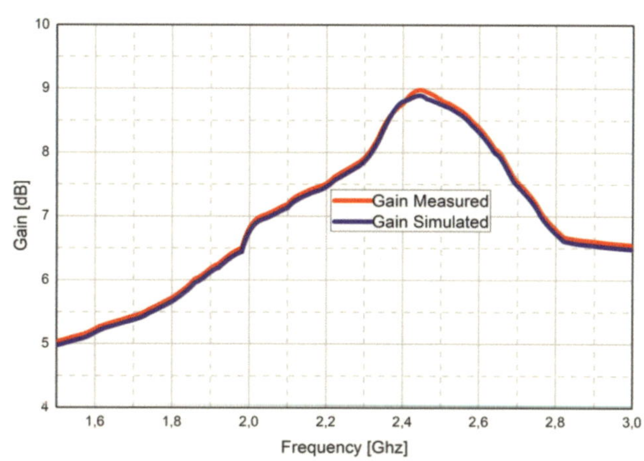

Figure 16.
Simulation and measurement gain using WCIP.

5. Conclusion

This chapter focuses on the design and production of antennas for radio frequency identification. This technology is experiencing a huge growth and requires the reduction of the cost of an electronic tag to meet the development of its market. Indeed, the reduction of the cost of an RFID tag is a crucial element for the unitary identification of goods. This cost must represent a negligible part compared to the product it identifies. An RFID tag is made of an electronic chip, a support and an antenna. To date, the manufacturing of RFID tags adopts the classic techniques of electronics. A solution for the reduction of the cost of an RFID tag could come from the use of

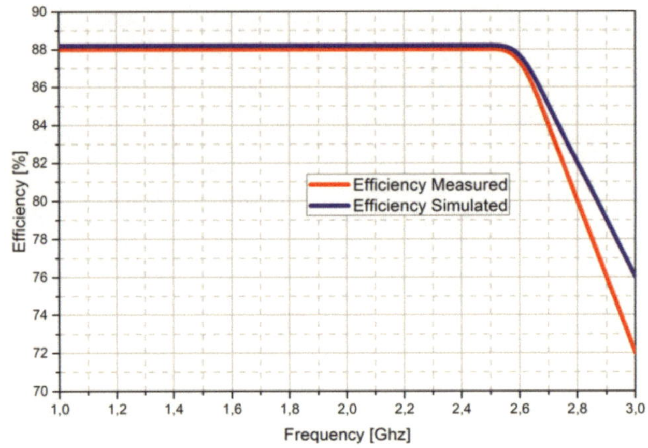

Figure 17.
Simulation and measurement efficiency using WCIP.

Parameter	Designation
λ	wavelength
P_r	power transmitted
G_r	reader antenna gain
G_a	tag antenna gain
d	distance between the tag and reader

Table 6.
Designation.

Parameter	Values
λ	0.12 m
P_r	1w
τ	0.92
G_a	8.96 dB
P_{th}	0.04 w
G_r	8 dB
r	9.58 m

Table 7.
The numerical values.

printing technologies. The design and realization of RFID tags, especially of their antennas, by using this innovative technology has been demonstrated. In this chapter, we succeeded in designing a miniaturized antenna that meets the requirements of RFID technology.

DOI: http://dx.doi.org/10.5772/intechopen.106990

Author details

Nizar Sghaier
Faculty of Sciences of Tunis, Department of Physics, Tunis EL Manar University, Tunis, Tunisia

*Address all correspondence to: niizar.sghaier@gmail.com

References

[1] Byondi FK, Chung Y. Longest-range UHF RFID sensor tag antenna for IoT applied for metal and non-metal objects. Sensors. 2019;**19**(24):5460. DOI: 10.3390/s19245460

[2] Naoui S, Latrach L, Gharsallah A. Metamaterials dipole antenna by using split ring resonators for RFID technology. Microwave and Optical Technology Letters. 2014;**56**(12): 2899-2903. DOI: 10.1002/mop.28731

[3] Ziai MA, Batchelor JC. Temporary on-skin passive UHF RFID transfer tag. IEEE Transactions on Antennas and Propagation. 2011;**59**(10):3565-3571. DOI: 10.1109/tap.2011.2163789

[4] Hoon WF, Malek MFBA, Fang LH, Seng LY, Zahid L. On the miniaturization high permittivity DRA with array patches. In: 1st International Conference on Artificial Intelligence, Modelling and Simulation, 03-05 December 2013. Kota Kinabalu, Malaysia: IEEE; 2013. pp. 443-445

[5] Lee B, Harackiewicz FJ. Miniature microstrip antenna with a partially filled high-permittivity substrate. IEEE Transactions on Antennas and Propagation. 2002;**50**(8):1160-1162. DOI: 10.1109/tap.2002.801360

[6] Shi Y, Whites KW. Miniaturization of helical antennas using dielectric loading. In: IEEE-APS Topical Conference on Antennas and Propagation in Wireless Communications (APWC); 03-09 August 2014. Palm Beach, Aruba: IEEE; 2014. pp. 163-166

[7] Porath R. Theory of miniaturized shorting-post microstrip antennas. IEEE Transactions on Antennas and Propagation. 2000;**48**(1):41-47. DOI: 10.1109/8.827384

[8] Ali T, Saadh MAW, Pathan S, Biradar RC. A miniaturized circularly polarized coaxial fed superstrate slot antenna for L-band application. Internet Technology Letters. 2018;**1**(6):e21. DOI: 10.1002/itl2.21

[9] Fan Z, Qiao S, Huang-Fu JT, Ran L-X. A miniaturized printed dipole antenna with V-Shaped ground for 2.45 GHz RFID readers. Progress in Electromagnetics Research. 2007;**71**:149-158. DOI: 10.2528/PIER07022501

[10] He S, Zhang Y, Li L, Lu Y, Zhang Y, Liu H. High performance UHF RFID tag antennas on liquid-filled bottles. Progress In Electromagnetics Research. 2019;**165**:83-92. DOI: 10.2528/PIER19041001

[11] Amin Y, Chen Q, Tenhunen H, Zheng L-R. Performance-optimized quadrate Bowtie RFID antennas for cost-effective and eco-friendly industrial applications. Progress In Electromagnetics Research. 2012;**126**:49-64. DOI: 10.2528/PIER12020805

[12] See KY, Chua EK, Liu Z. Accurate and efficient evaluation of MoM matrix based on a generalized analytical approach. Progress in Electromagnetics Research. 2009;**94**:367-382. DOI: 10.2528/PIER09063002

[13] Jing Y-F, Carpentieri B, Huang T-Z. Experiments with Lanczos Biconjugate a-orthonormalization methods for MoM discretizations of Maxwell's Equations. Progress In Electromagnetics Research. 2009;**99**:427-451. DOI: 10.2528/PIER09101901

[14] Chen M, Zhao X-W, Zhang Y, Liang C-H. Analysis of antenna around nurbs surface with iterative Mom-Po technique. Journal of Electromagnetic Waves and Applications. 2006;**20**(12):

1667-1680. DOI: 10.1163/156939306779292372

[15] De la Rubia V, Zapata J. A methodology for antenna design via the finite element method. In: 2nd European Conference on Antennas and Propagation (EuCAP); 11-16 November 2007. Edinburgh: IET; 2007. pp. 1-10

[16] Sheng X-Q, Yung EKN. Analysis of microstrip antennas on finite chiral substrates. International Journal of RF and Microwave Computer-Aided Engineering. 2003;**14**(1):49-56. DOI: 10.1002/mmce.10115

[17] Dash JC, Jena MR, Mangaraj BB. Analysis of Dipole Antenna and its array using finite element method. 2016 2nd International Conference on Advances in Electrical, Electronics, Information, Communication and Bio-Informatics (AEEICB). 2016. DOI: 10.1109/aeeicb.2016.7538326

[18] Tirkas PA, Balanis CA. Finite-difference time-domain method for antenna radiation. IEEE Transactions on Antennas and Propagation. 1992;**40**(3):334-340. DOI: 10.1109/8.135478

[19] Lampe B, Holliger K, Green AG. A finite-difference time-domain simulation tool for ground-penetrating radar antennas. Geophysics. 2003;**68**(3):971-987. DOI: 10.1190/1.1581069

[20] Houaneb Z, Zairi H, Gharsallah A, Baudrand H. Modeling of cylindrical resonators by wave concept iterative process in cylindrical coordinates. International Journal of Numerical Modelling: Electronic Networks, Devices and Fields. 2011;**24**(2):123-131. DOI: 10.1002/jnm.765

[21] Gharsallah A, Mami A, Douma R, Gharbi A, Baudrand H. Analysis of a microstrip antenna with fractal multilayer substrate using iterative method. International Journal of RF and Microwave Computer-Aided Engineering. 2001;**11**(4):212-218. DOI: 10.1002/mmce.1026

[22] Richalot E, Wong MF, Baudrand H, Fouad-Hanna V. An iterative method for modeling of antennas. International Journal of RF and Microwave Computer-Aided Engineering. 2001;**11**(4):194-201. DOI: 10.1002/mmce.1024

[23] Djouimaa A, Titaouine M, Adoui I, et al. Tunable FSS simulation using WCIP method for multiband and dual polarized applications. Radioelectronics and Communications Systems. 2017;**60**:106-112. DOI: 10.3103/S0735272717030025

[24] Cassivi Y, Wu K. Substrate integrated nonradiative dielectric waveguide. IEEE Microwave and Wireless Components Letters. March 2004;**14**(3):89-91. DOI: 10.1109/LMWC.2004.824808

[25] Baudrand H, Titaouine M, Raveu N, Fontgland G. Electromagnetic modeling of planar almost periodic structures. In: SBMO/IEEE MTT-S International Microwave and Optoelectronics Conference (IMOC); 03-06 November 2009. Belem, Brazil: IEEE; 2009. pp. 427-431

[26] Ammar N, Baudrand H. Wave concept iterative process method for multiple loop antennas around a spherical media. IET Microwaves, Antennas and Propagation. 2019;**13**:666-674. DOI: 10.1049/iet-map.2018.5661

[27] Hlali A, Houaneb Z, Zairi H. Dual-band reconfigurable graphene-based patch antenna in terahertz band: Design, analysis and modeling using WCIP

method. Progress in Electromagnetics Research C. 2018;**87**:213-226. DOI: 10.2528/PIERC18080107

[28] Liu Z-L, Wang C-F. Mutual coupling analysis of multiple on-board antennas with sub-domain MoM-PO method. In: IEEE International Symposium on Electromagnetic Compatibility and 2018 IEEE Asia-Pacific Symposium on Electromagnetic Compatibility (EMC/APEMC); 14-18 May 2018. Suntec City, Singapore: IEEE; 2018. p. 89

[29] Latifa NB, Aguili T. Synthesis and optimization of almost periodic antennas using Floquet modal analysis and MoM-GEC method. Journal of Electromagnetic Analysis and Applications. 2019;**11**:1-16. DOI: 10.4236/jemaa.2019.111001

[30] Pereira-Filho OMC, Barbosa HB, Diniz CA, Tinoco-S AF. Method of moments analysis of arrays of cylindrical microstrip antennas with superstrate. IET Microwaves, Antennas and Propagation. 2018;**12**(4):561-568. DOI: 10.1049/iet-map.2016.0986

[31] Shi Y, Qi K, Liang C-H. A miniaturized design of 2.45-GHz RFID tag antenna. Microwave and Optical Technology Letters. 2015;**57**(8): 1905-1908. DOI: 10.1002/mop.29223